Rome's Armies to the Death of Augustus

Rome's Armies to the Death of Augustus

Tony McArthur

Pen & Sword
MILITARY

First published in Great Britain in 2024 by
Pen & Sword Military
An imprint of Pen & Sword Books Limited
Yorkshire – Philadelphia

Copyright © Tony McArthur 2024

ISBN 978 1 39908 007 1

The right of Tony McArthur to be identified as
Author of this Work has been asserted by him in accordance
with the Copyright, Designs and Patents Act 1988.

A CIP catalogue record for this book is
available from the British Library

All rights reserved. No part of this book may be reproduced or
transmitted in any form or by any means, electronic or mechanical
including photocopying, recording or by any information storage and
retrieval system, without permission from the Publisher in writing.

Typeset by Mac Style
Printed in the UK by CPI Group (UK) Ltd, Croydon, CR0 4YY.

Pen & Sword Books Limited incorporates the imprints of After
the Battle, Atlas, Archaeology, Aviation, Discovery, Family History,
Fiction, History, Maritime, Military, Military Classics, Politics,
Select, Transport, True Crime, Air World, Frontline Publishing, Leo
Cooper, Remember When, Seaforth Publishing, The Praetorian Press,
Wharncliffe Local History, Wharncliffe Transport, Wharncliffe True
Crime and White Owl.

For a complete list of Pen & Sword titles please contact

PEN & SWORD BOOKS LIMITED
47 Church Street, Barnsley, South Yorkshire, S70 2AS, England
E-mail: enquiries@pen-and-sword.co.uk
Website: www.pen-and-sword.co.uk
or
PEN AND SWORD BOOKS
1950 Lawrence Rd, Havertown, PA 19083, USA
E-mail: uspen-and-sword@casematepublishers.com
Website: www.penandswordbooks.com

To my family, particularly to my wife of almost fifty years, Patricia, whom I dragged over the remains of Maiden Castle on our honeymoon! They have patiently listened to my historical obsessions, in the case of my four children Andrew, Catherine, James and Elizabeth, since they were born. This volume is dedicated to them all.

Contents

List of Plates — ix
Introduction — xi

Part I: Rome's Armies and Historical Assumptions — 1

Chapter 1   The History of the Study of the Romans' Armies — 3

Chapter 2   The Roman Army? — 24

Chapter 3   Changes in Rome's Armies, the Problem with 'Reform' and an Alternative — 32

Chapter 4   The Myth of Professionalism in Rome's Armies before 14 CE — 56

Part II: Rome's Armies Before and After Polybius — 71

Chapter 5   Earliest Roman Armies — 73

Chapter 6   Polybius and his Roman Armies — 86

Chapter 7   Rome's Armies in Caesar's Time — 110

Chapter 8   The Armies of Imperator Augustus — 132

Conclusion — 155
Abbreviations — 157
Notes — 161
Bibliography — 205
Index — 222

List of Plates

Figure 1. Niccolò di Bernardo dei Machiavelli (1469–1527) by Santi di Tito. (*Wikimedia Commons/public domain*)

Figure 2. Raimondo Montecuccoli (1609–1680) by Elias Grießler (1622–1682). (*Wikimedia Commons/public domain*)

Figure 3. Jean Charles, Chevalier de Folard (1669–1752). (*Millearia via Wikimedia Commons/CC BY 3.0 DEED*)

Figure 4. Maurice de Saxe, i.e. Hermann Moritz von Sachsen (1696–1750) by Maurice Quentin de La Tour. (*Wikimedia Commons/public domain*)

Figure 5. Antoine-Henri Jomini (1779–1869) by workshop of George Dawe. (*Wikimedia Commons/public domain*)

Figure 6. Hans Gottlieb Leopold Delbrück (1848–1929). (*Wikimedia Commons/public domain*)

Figure 7. Certosa Situla, from Etruscan antiquities in the Archeological Museum of Bologna, with illustrations of figures with different equipment in the top row. (*Sailko via Wikimedia Commons/CC BY-SA 3.0 DEED*)

Figure 8. The front cover of von François's *Tannenberg*. (*The Internet Archive*)

Figure 9. Parade of the 'Prussian' 1st Foot Guard Regiment in the Lustgarten, 9 February 1894 (Potsdam) by Carl Röchling. (*Wikimedia Commons*)

Figure 10. South Italian bronze helmet from the third century BCE. (*The British Museum*)

Figure 11. Section of the Pydna Monument showing two helmets. Note the centre grip in the large shield of the figure in the right. The cavalryman's armour is in a Greek style, supporting Polybius' statement that the Romans adopted Greek style armour for their cavalry. (*Colin Whiting via Wikimedia Commons/CC BY-SA 4.0 DEED*)

Figure 12. Roman soldiers on the monument known as the Altar of Domitius Ahenobarbus. (*Wikimedia Commons/public domain*)

Figure 13. Commemorative stele of Centurion M. Caelius who died in Varus' forces in Germany. (*Wikimedia Commons/public domain*)

Figure 14. Sketch of a monument from Urso, Spain, showing Roman soldiers, Republican period from the shield shape. Note that the left figure appears to be wearing armour but the figure on the right is depicted in cloth only. Both appear to be wearing greaves.

Figure 15. Sketch of the funeral stele of Centurion Minucius, Legio Martia in Padova Musei Civici. He has his military belt with a *pugio* (dagger). He carries his vine stick, a symbol of office. Note that he is not shown wearing armour.

Figure 16. Modern re-enactors in France, at Vienne, in 2015.

Introduction

There is certainly no shortage of books on what is commonly termed, the 'Roman army' but, strictly speaking, this book is not about the 'Roman army' but Rome's armies. The distinction is argued later but for the moment, the difference between the two terms is that 'Rome's armies' does not refer to a social institution like national armies in modern states, implied by the term 'Roman army'. This distinction, like a number in this book, explains why the book has been written. In 2023, such has been the accretion of assumption and accepted views on the Romans' military institutions, it is necessary to review what serves as the accepted wisdom on Rome's armies from earliest times to the death of Augustus Caesar in 14 CE.

Rome's armies have never been written about strictly for themselves. The armies have been of interest because of what they meant to and for others. The earliest surviving references to Rome's armies in Polybius refer to them because of what he believed they meant to his Greek readers who struggled to understand how the Macedonian armies could have been defeated, and later Josephus did the same for his readers after the death of Augustus. Modern authors, who write about Rome's armies, do so because of what they mean today for military or political history. Prior to the mid-nineteenth century, Rome's armies were of interest for what could be learnt for contemporary military operations. The many references to Rome's armies in works dedicated to Roman history simply see the armies as an aspect or an artefact of political and social developments. This work aims for something of a 'reset' in that it attempts to confine itself to what is known about the armies and, perhaps more importantly, what is not.

In part, this work is an historiographical study of the Romans' military institutions. Jean-François Lyotard advocated the questioning of metanarratives almost fifty years ago, but some scholars remain uncomfortable with implications of literary theory for history.[1] This is quite ironic given that since Thucydides' and Polybius' criticisms of other writers, historians have argued about how history should be written.[2] The study of Rome's military institutions suffers from an unquestioned acceptance of assumptions which support metanarratives. These assumptions include concepts like 'reform' to characterize change, or that the Romans had 'standing armies' prior to Augustus. Much responsibility should be sheeted home to the metaphors we accept, often unconsciously, and even more to the ones we reject. The common view that the Romans had a 'modern' army, with whatever that implied, has limited our understanding of how very different the Romans and their institutions were compared with our world.

In many ways, this book will have to emphasize more of what we don't know than what we think that we do. What we know can be likened to shining a narrow beam of light on a very large, darkened wall, showing only small spots. For this reason, some will find what has been referred to as the 'nuts and bolts' approach to military history unsatisfying. But like Hans Delbrück, I am guided by the sources.[3] What Polybius did for the second century BCE, the self-serving Caesar for the first century BCE, the bitter Tacitus for the first century CE, Arrian for one campaign in the second century and disillusioned Ammianus for the fourth tells us explicitly are these narrow beams. Their shadowy edges compose the occasional mentions in other works and the archaeological finds, all of which must be contextualized within narratives, having no voice of their own. We can add a variety of monuments and inscriptions, some grand like Trajan's propagandist column and some artistically naive but more trustworthy like the Adamklissi metopes. From these trustworthy scattered sources and others less so, we can build a very partial picture of Rome's armies. But for this work, we shall look only at the light these texts can shed on Roman armies up to 14 CE.

This work aims to both describe and explain how and why Rome's armies changed from our earliest knowledge of them to the death of

Augustus. I will avoid descriptions of campaigns and battles unless they are germane to this focus - there is little point in retelling Livy's, Polybius' and Caesar's stories. This end point of 14 CE has not been chosen for convenience, or because of the common division of Roman history between republic and empire. As will be discussed later, an unintended, coincidental result of Augustus' long life and domination of Roman politics was the 'undying' army. Unlike all previous armies commanded by Roman magistrates, Augustus' command did not end and, as a result, his forces were not disbanded. Further, because his command was shared with his adoptive son, Tiberius, his command extended beyond his death. The armies which Tiberius continued to command became the Imperial armies, but that is a matter for another day.

In the pages that follow, I will attempt to deconstruct two key aspects of current views of the Romans' military arrangements to the death of Augustus. I have already referred to the first: the idea that the 'Roman army' existed in an institutional sense. The second is how changes in Rome's armies can be explained across time by occasional 'reforms'. It is, however, insufficient to merely deconstruct: it is also necessary to offer an alternative, replacement view. The nature of Rome's armies will be examined in terms of contemporary accounts, principally Polybius', and Caesar's. What these will show is the capacity of Roman generals for contingent adaptation to their circumstances. Unlike modern commanders who work out of doctrine and Standard Operating Procedures (SOPs), ancient generals were guided by customary practices which they adapted to the circumstances in which they operated. What emerges is a power whose large population and resilient institutions allowed it to militarily dominate the Mediterranean world. This should not surprise us. Victory in the two great wars of the twentieth century was achieved by the *Materialschlacht*, i.e., resources, war materials and population battle, in which the USA and the USSR were far superior to the Central and Axis powers.

What is commonly termed the 'Roman army' itself has a history. Today, it is almost solely studied for historical interest so it seems quite odd to us to imagine that eminent generals of the seventeenth and eighteenth centuries would have studied Roman tactics, as they understood them, to inform their contemporary military practice. However, if they knew how we studied Rome's armies, these same generals would be equally perplexed because Roman military practices and experience are of little relevance to modern soldiers. As we recognize the flawed discourses of the past, it is sobering to know that scholars in the future are very likely to wonder at our naivety and errors. Equally, we need to have the humility to accept that the scholars who preceded us believed that they were improving our understanding of the past and would be surprised at how we view it. Nevertheless, we should have the confidence to accept their different perspectives while being governed by our own view of the Roman past, in a manner which is consistent with our times. The study of history is a continual conversation between the present and the past. We may admire Thucydides, Polybius, Tacitus and Ammianus but being motivated to describe one's own times is considerably closer than most would want to the Chinese curse: 'better to be a dog in times of tranquillity than a human in times of chaos' (寧為太平犬, 不做亂世). To engage in the distant past is a rare luxury to be treasured. To study the military activities of an ancient Mediterranean power for interest is surely a sign of the mostly unearned prosperity enjoyed only by the wealthiest in the Romans' world.

This work is not aimed exclusively at academic scholars. Mass education has enabled a vast number of people to enjoy what was once the preserve of a privileged few. I will be quite happy if academically employed scholars find value in this work, but they are not the primary audience. I would like people with an interest in Rome's armies to read this work critically. I do not seek agreement; I seek engagement with the ideas. If I am wrong, then show the vast community of interested persons why. To do so is to advance our collective understanding.

Tony McArthur
Canberra
April 2023

Part I

Rome's Armies and Historical Assumptions

Chapter 1

The History of the Study of the Romans' Armies

Introduction

The 'Roman Army', and its cognates in different languages, is a frequently encountered term for Rome's military forces. Scarcely a month passes without the appearance of a book or article with the term in its title.[1] The continuing availability of works on the Roman Army reflects a healthy publishing market, based on popular interest. One respected scholar described a common, if mistaken, view: 'Our image of the Roman army is one of massive organization, rigidly disciplined might, and, most of all, incredible modernity'.[2] While this characterization of the Romans' armies may reflect a common view, it is a comparatively recent one. Interest in the Romans' military institutions can be traced from the ancient world. Initially, interest lay in learning from the Romans how to be successful in war; later, antiquarian enthusiasm evolved into scholarly research, but the idea that contemporary military practice could learn from the Romans has not disappeared. Our modern 'Roman Army' is a comparatively recent discovery, but did it exist before 14 CE?

Rome's Armies Prior to the Mid-Nineteenth Century

People have studied the Romans' armies for over two millennia. The different authors who have written about them have done so to address issues contemporary to them, some as soldiers like Arrian but others have written about them to learn or teach others. Some of their works are classed as military manuals, but others are histories.[3] The earliest surviving extended text we have which explicitly describes the Romans' military institutions is found in Book VI of Polybius' history. Polybius' purpose in writing his work was to explain Rome's success as an imperialist power to his Greek speaking audience.⁴ In Book IV, he provides an explanation in the context of what he terms Rome's πολιτεία (*politeia*), a word commonly translated as 'constitution' but with a broader meaning in Greek, referring to the society and its customs.⁵ He describes the Romans' military arrangements in detail, noting their willingness to adopt other nation's practices.⁶ In Book XVIII, Polybius explains why the Romans equipment and tactics were superior to the Macedonian phalanx, well-known to his Greek audience

For Romans after the death of Augustus, the historian Livy provided a detailed narrative of Rome's history with a particular concentration on her military experience. Writing at the time of Augustus, he described Rome's history, and, like any good historian, he had a narrative purpose. Until we are served by Polybius, who is mostly thought to be more reliable, Livy, with some support from Dionysius of Halicarnassus, is our guide to the Romans' earliest military institutions.⁷ Livy's intent is defined in his preface: *qui mores fuerint, per quos viros quibusque artibus domi militiaeque et partum et auctum imperium sit.* i.e., "what morals were like through what men and by what policies, in peace and in war, empire was established and enlarged ".⁸ To Livy, domestic and military affairs must both be considered to explain the empire. Livy uses his narrative to convey his purpose: to display by what means (*quibus artibus*), both domestically and militarily (*domi militiaeque*), the empire (*imperium*), came into being and increased.⁹ Unlike Polybius, Livy believes that moral qualities, not the institutions, determined the fate of the empire.¹⁰ His detailed description of the *devotio* of Decius Mus at the Battle of Sentinum in 295 BCE presents a moral exemplum for Roman citizens at war.¹¹ The Roman defeat at the Allia in 387 BCE is explained as a moral failure by the Romans to maintain their devotion to the gods.¹² Livy had a coherent sense of war but did not see the details of organization and manoeuvre as the reasons for Roman victories.¹³ Livy can be read to discern organizational and even tactical detail but such things are coincidental to his view of Roman military success or failure.¹⁴ Four centuries later, Publius Vegetius Renatus' *de re Militari*, a very different text to those of Livy and Polybius, was produced. Unlike them, it is not a history but what we call a 'military manual'. It appears to advise on the recruitment and training of Roman soldiers, based on past Roman experience.[15] Vegetius advocates a return to earlier practices which he believes were effective, with the past providing solutions for the present. The Eastern Romans, whom we call Byzantines, but who called themselves Romans, had always studied their ancient military past as an aid to their present.[16] Scholars today struggle with distinguishing Byzantine material from that drawn from Greek and earlier Roman texts.[17] For the Byzantines there was no reason to distinguish between the two because Greek or Roman texts were intrinsic to their cultural identity. The *Strategikon*, attributed to the Emperor Mauricius (582–602 CE), but perhaps written by others, on the organization of the army, refers to the legions and to a ἀριθμός i.e., *numerus*, a Latin term of a military unit in the Late Empire, both terms used in the past.[18] There are quite a number of later Byzantine military texts which borrow heavily from Greek and Roman authors.[19] Their authors' aims were to address concerns in their times.[20]

Between the sixteenth and the nineteenth centuries, most eminent European military theorists and practitioners studied the armies of the Romans for ideas to address contemporary military issues. Latin texts were available prior to the Renaissance. Vegetius was read for a number of purposes and translated but there was little interest in the Romans' military as opposed to Vegetius' moral admonitions.[21] With the more ready availability of classical works, particularly Greek works following the fall of Constantinople in 1453, and the invention of the printing press in Europe, Europeans interested in warfare, copying the Byzantines, began to study Roman military experience more closely from the sixteenth century.[22] Renaissance humanist interest in Latin historical texts is reflected in the writing of Biondo Flavio, a senior member of the Papal Curia who died in 1463. Biondo's *Roma Triumphans* was an antiquarian collection using the headings *antiquitates, publicae, sacrae, militaries*.[23] Under *militaries*, Biondo compared contemporary armies unfavourably to Roman armies.[24] Justus Lipsius (1547–1606), advocating for improved armies in his time, produced a work on the Roman military using Polybius.[25] The Dutch found military forebears in the Batavian *auxilia* who revolted from Rome, as the Dutch had rebelled from Spain.[26] Contrary to modern interests, the aim of these authors was primarily to improve the success of their own and their ruler's armies. Niccolò Machiavelli (see Figure 1) was a Florentine statesman who is most well-known for his 1513 work *Il Principe* but his interests were not confined to what is referred to as 'politics' today. Machiavelli's *Dell'arte della guerra*, was an influential text in Renaissance Europe in which he tried to bring the military experience of the past, including that of the Romans, to bear on contemporary issues.[27] First published in 1521 in Italy, his work was translated into Spanish, French, English, German and Latin.[28] Machiavelli's approach to improving the military arrangements in Florence was underpinned by the assumption that the ancient past provided the appropriate starting point. As an educated man of his time and place, Machiavelli was familiar with Livy about whom he wrote.[29] Translations and texts of Frontinus, Aelian, Vegetius and Modestus were available in Machiavelli's time.[30] Based on his comments at the beginning of Part 2, he may have been familiar with Polybius; his mentions of Tiberius and Augustus suggest a knowledge of Tacitus. Machiavelli's recommendations for improved military arrangements for his own time assume that his readers will view ancient precedents as the basis for improvements in their time. For Machiavelli, the lack of discipline of contemporary armies and their lack of organization induced him to seek remedies in the past.[31] In terms of his own time, Machiavelli was particularly impressed with the *virtù* that he saw in the Romans.[32] His work on war in the form of a conversation explicitly cites the armies of the Romans as models for his time. At the same time, it is quite clear that his ideas are rooted in contemporary concerns when he compares the German (Imperial) and Swiss troops with Roman practices.[33] Contrary to modern assumptions, Machiavelli does not refer to the Roman army in any institutional sense. His focus is on the moral attractiveness of Roman institutions, perhaps echoing Polybius, and the resulting military arrangements although he does not ignore changes which he believed reduced the effectiveness of Roman armies.[34]

The emergence of modern European states through what are commonly known as the 'Wars of Religion', between the Knight's Revolt in 1522 and the Peace of Westphalia in 1648, saw considerable changes in military practices, particularly in the use of gunpowder weapons. The works of Raimondo Montecuccoli (see Figure 2), born in Italy in 1609, exemplify a rich literary tradition which sought to subject warfare to contemporary theories of knowledge, some of which are decidedly odd to modern eyes.[35] He began his military career in 1625 in the service of the Holy Roman Emperor. He served with great distinction as a senior officer in the Imperial armies against the French, the Ottomans and the Empire's northern enemies. His memoirs are not the product of armchair theory or a limited experience of war, as Machiavelli's views may have been; he was among the best generals of his day. Yet even this great commander, who is compared to Fabius Maximus by contemporaries, underpins his writing by reference to the Greeks and Romans.[36] Montecuccoli states that: 'The Greeks and Romans left us fine examples of the choice and arrangement of soldiers' (*Les Grecs et les Romains nous ont laissé de beaux examples du choix et de l'arrangement des soldats*). Montecuccoli assumes that the military experience of the Greeks and Romans were relevant to the conduct of war in his day.[37] He makes frequent reference to Roman practices.[38] Montecuccoli assumes that his readers will be so familiar with ancient practices that further elucidation was not required.[39] This is underlined by how he compares, in passing, the manner in which the Romans organized their troops and the *Ordannances militaires de l'Empereur Charles V*.[40] However Montecuccoli is not ignorant of the impact of gunpowder and the degree to which it limits the reliance which can be placed on ancient precedents.[41]

Leading early-eighteenth-century soldiers in Europe continued to rely on ancient military experience. The Chevalier de Folard (see Figure 3) was a successful and highly esteemed soldier of the first half of the eighteenth century. He served in the armies of the French and Swedish kings. He was particularly interested in Polybius and undertook to provide a commentary upon retirement from active service.[42] For de Folard, the ancient texts he studied were of immediate relevance to the military problems of his day and he put the texts to service in those endeavours. De Folard wished to promote military tactics which were more suitable to the nature of French people rather than simply copying what worked in the armies of the King of Prussia, for example.[43] In the contemporary debate on the relative merits of the use of the line and the column in infantry tactics, de Folard argued strongly, based on ancient evidence, that the column was better for an attack than a line.[44] Maurice de Saxe (see Figure 4), one of the many extramarital children of Augustus II, King of Poland and Elector of Saxony, had a distinguished military career. He served the Russian Emperor, the Holy Roman Emperor and finally the King of France under whom he became a marshal, winning a famous victory over the British at Fontenoy in 1745. De Saxe is the posthumous author of an unusual work known as *Mes Rêveries*. This document is heavily dependent upon Roman military practice in particular, although the Greeks' experience is not ignored: 'We must have a better opinion of the ancients and the Romans, who are our masters, or who should be' (*Il faut avoir mellierure opinion des anciens et des Romains, qui sont nos maîtres, ou qui devraient l'être*).[45] Reflecting his own day, de Saxe assumed that the Romans marched *en cadence*, i.e. in step, showing that the dangers of assumption are not confined to our time.[46] As the matter of the cadenced step suggests, the purpose of de Saxe's *Mes Rêveries* was not antiquarian. De Saxe was serious about challenging the military practices of his time and he used the respect in which Greek and Roman military practices were seen by his contemporaries to promote his ideas. Stressing that Roman military methods were superior to contemporary practices, he dismisses concerns that the Romans lacked gunpowder.[47] De Saxe claims that contemporary firearms are not as effective as some imagine, a claim born out to a degree by modern research.[48] De Saxe devoted considerable effort advocating for the use of Roman organization for contemporary armies.[49] In one respect this was quite prescient. He notes that Roman legions contained both cavalry and infantry whereas contemporary practice was to divide them.⁵⁰ Only forty years later the French began to introduce the corps system of combined arms which was central to the effectiveness of French armies during the Napoleonic wars. De Saxe should not be seen as rooted in the past although he saw no virtue in novelty.⁵¹ He recognized that some contemporary practices were considerably superior to ancient equivalents; he singles out fortifications in particular.⁵²

There was a an enormous increase in the writing on military matters in Europe from the middle of the eighteenth century, a period often known as the Eighteenth-Century European Enlightenment.⁵³ During this time, reference was typically made to antiquity in works on military affairs by authors most esteemed by contemporaries.⁵⁴ The *Encyclopédie* which appeared in 1751, edited by Diderot and D'Alembert, in its entry on *Guerre* (war), argued that theory of war was the result of the combination of knowledge from ancient authors, further developed from more recent experience.⁵⁵ The French soldier, Paul-Gédéon Joly de Maïzeroy (1719–1780), in his two-volume work on tactics, refers to the Romans 200 times. He was an able classicist as well as an accomplished officer. His military works were widely read in his time. He believed that the Romans and the Greeks could be models, not only in the arts, but also in military affairs.⁵⁶ He acknowledged the changes since the sixteenth century as a 'révolution' but still believed that the ancients provided guidance.⁵⁷ He believed that while gun powder had required changes in military operations, the bases of tactics and grand manoeuvres had not changed, arguing for shock over firepower tactics.⁵⁸ Karl Gottlieb Guischardt, born in Holland, a son of French refugees, was in the service of Frederick the Great. He was an avid student of ancient military history and the commander of Frei Korps units in the army of the Prussian king. Guischardt, like his contemporaries, believed that, regardless of the discovery of gun powder 'the military art of the ancients will always be the school of good officers' (*l'art militaire des anciens sera toujours l'école des bons officiers*).⁵⁹ Guischardt's work used ancient texts to deduce lessons for officers in his own time.⁶⁰ He was familiar with some close contemporary writers and their views of the ancient texts. He engages with the French Marshal Puysegur's principles in relation to the value of training.[61] He discusses the debate between de Folard and de Saxe regarding the use of columns in attack.[62] Guischardt devotes quite some time to the differences between the Macedonians and the Romans to deduce lessons for his time. He even refers to Roman arrangements using the contemporary French term *Ordonnonce*, i.e. 'regulation'.[63] He contrasts how the Romans charged compared to 'our way of charging' (*notre manière de charger*).[64] While valuing the experience of the Greeks and Romans, he, like de Saxe, acknowledges the limitations of using them as a source for the defence of fortifications, noting that gunpowder has made ancient walled defences out of date.[65] Not all military writers in this period were as dependent upon ancient views. Jacques-Antoine-Hippolyte, Comte de Guibert was an officer in the army of the King of France. His essay *Essai général de tactique*, published in 1770 in London, was widely admired and translated into other languages. While his essay addresses contemporary issues, his work shows a deep engagement with military history including that of the ancient world. De Guibert appreciated that armies of his day are both larger and used firepower unknown to ancient generals who appeared to have known little of topography.[66] However he did not believe that the military arts had improved on those mastered by Greek and Roman generals. He compared Frederick II of Prussia to Alexander the Great and the Austrian Marshal Daun to the Roman opponent of Hannibal, M. Fabius.[67] While being aware of the changes which occurred in the ancient world, he sees the Romans as a model for discipline.[68] Reflecting on the work of earlier writers, the Prussian soldier von Clausewitz saw Maïzeroy and Guischardt as the most expert on the art of war in antiquity.[69]

The strict rationalism of the Eighteenth-Century Enlightenment, particularly in its French application, was not the only significant intellectual movement for military theorists toward the end of the century. That there should be a consistency between philosophical and military ideas developed at the same time should not surprise us. Military writers are subject to the same intellectual forces as others who study different domains of knowledge. There were two particularly important intellectual developments which appeared in the latter half of the eighteenth century: the reaction to rationalism and historicism. Kant's *Kritik der reinen Vernunft* (Critique of Pure Reason), first published in 1781, inspired those who saw French rationalism as somewhat barren and lacking an appreciation of the complexities of contemporary institutions. Its publication in German, rather than French, the common language of European intellectuals, is significant. This work was highly influential, particularly in the German-speaking lands. The second development was historicism, which looked to understand how situations in the present could be seen through understanding how the past had shaped the present.[70] This approach should be seen as distinct from history as a discipline which aims, instead, to understand the past more in its own terms. The Hegelian world view, which explained the Prussian kingdom as a reflection of the evolution of appropriate human freedom through the operation of a spirit of the age, is perhaps the best example of the application of historicism. These two developments, when applied to the experience of the Napoleonic and Revolutionary wars, changed how military affairs were seen in Europe, particularly in Germany. Both combined to provide a critique of rationalism and an encouragement to those who were attracted to traditional practices and institutions, even to nationalist sentiments in Germany. While there was little impact on military affairs initially, towards the end of the eighteenth century European states, including the German-speaking states, began to establish military schools to train officers, principally.[71] Recommended reading included the military experience of antiquity but other ideas were important too.[72] The illegitimate son of Leopold I of Anhalt-Dessau, Georg Heinrich von Berenhorst (1733–1814) who served in the Prussian army under Frederick the Great, was the author of an influential work on warfare.[73] He, like most military writers in this period, accepted that Graeco-Roman military experience was still relevant but, significantly, argued that contemporary European practices had developed from the Renaissance to the expertise demonstrated by Frederick the Great. However, Berenhorst was highly critical of Frederick's almost-mechanical, rationalist approach because it ignored the potential of moral commitment of the troops themselves.[74] Berenhorst's ideas were taken further by a group of German writers, described as military *Aufklärers* (enlighteners), who began to question the relevance of ancient military experience.

The wars fought in Europe between the French Revolution and the defeat of Napoleon at Waterloo in 1815 not only changed the political face of Europe, they changed the way Europeans waged war. At the simplest level, European armies in this period were larger than ever before and they no longer depended upon the valuable, highly trained professional soldiers of the eighteenth century. Nothing symbolized this change more than the destruction of the Prussian army, still trained in the style of Frederick the Great, by Napoleon's French troops in the Battle of Jena in 1806. Following the wars of Napoleon in Europe, direct application of the military experience of the Greeks and the Romans lessened; however, they were still considered relevant and a useful source of military wisdom although of less use tactically. Antoine-Henri, Baron de Jomini (see Figure 5) was an officer in Swiss, French and Russian service between 1798 and 1859. His writings were particularly influential in the USA.[75] De Jomini exemplifies the impact of the wars of the Napoleonic period on European military thinking. His *Traité des grandes opérations militaires, contenant l'histoire critique des campagnes de Frédéric II, comparées à celles de l'empereur Napoléon: avec un recueil des principes généraux de l'art de la guerre* references Caesar, the Emperor Julian and Alexander the Great. Like his predecessors, de Jomini in his many works compares European military art in his time to ancient standards in terms of generals in Greek and Roman history.[76] De Jomini presents Frederick the Great as 'a new Caesar' (*un nouveau César*).[77] For de Jomini, the lessons of the past can be learned from the ancient past as well as, for him, more recent times.[78] Discussing strategy, he declares that the principles remain the same in his day as they were to the Scipios, Caesar, Frederick the Great, Peter the Great and Napoleon.[79] De Jomini refers to the Romans fleetingly although he does acknowledge that, in relation to the composition of armies and of officers, Roman institutions are worthy of attention.[80] In a discussion of the use of divisions in the

The History of the Study of the Romans' Armies 13

French Revolutionary armies, he compares these with Roman legions.[81] He was familiar with Vegetius, referring to the Romans' unwillingness to serve in the armies and Roman arrangements for magazines.[82] But not all agreed with De Jomini's use of ancient military experience as a point of reference. Drawing on his experience of the Napoleonic wars, the Prussian military writer, von Clausewitz declared, based on his own experience, that ancient military experiences was of little use to contemporary soldiers.[83] Slowly, writers on military affairs, while still expressing appreciation for Roman and Greek military experience, began to concentrate on the differences between ancient and contemporary military realities.

A more common view was found in an influential, multi volume textbook, published from 1833, on cavalry, infantry and artillery tactics, produced by Karl von Decker, a colonel in Prussian service and a member of the Swedish Royal Academy for Military Sciences. The work reflected Prussian practice but was considered to have a sufficiently wide application to be translated into French. Von Decker believed that Roman tactics (*römischen Taktik*) remained a topic of great interest and Roman military institutions could still serve as 'mother' to armies of his day.[84] Jean-Thomas Rocquancourt, a former Captain of Engineers in the French Army, became an instructor and Premier Sous-Directeur des Études Militaires at the École Militaire at St Cyr in 1812. Rocquancourt eventually published his course of instruction in three volumes in Brussels and Paris between 1837 and 1840. Rocquancourt continued the practice of earlier European scholars in military affairs in using ancient military experience, as transmitted by ancient authors, as a basis for contemporary military practice. Commanders in early-modern Europe are presented as heavily engaged with ancient military practices to inform European military practices.[85] In preparing his students for military command, he makes reference to changes in military organization and technology with frequent reference to the Greeks and the Romans. He notes the still-ongoing tactical debate regarding depth and breadth with reference to both Greeks and Romans.[86] The need in his day for reliance on magazines to supply armies is contrasted with Greek and Roman practices.[87] Rocquancourt, agreeing with de Saxe and Guischardt, argued against those who believed that the invention of gunpowder made the military experience of the Greeks and the Romans irrelevant.[88] He does however acknowledge that more recent military experience is more relevant because ancient tactics were simpler, armies smaller and spheres of action less extended.[89] Rocquancourt discusses both Greek and Roman military organization, at one point seeing the battalions of his day as the equivalent of Greek and Roman formations.[90] The contemporary military challenge of 'passage of lines', moving formations through other formations close to the enemy, is discussed in reference to the battle of Zama (202 BCE) between the Romans and Carthaginians.[91]

The Marquis de Carrion-Nisas (1767–1841) was an experienced French officer who commanded troops in Napoleon's armies. His two-volume work of 1824, *L'Essai sur l'histoire générale de l'art militaire* makes repeated references to both Greek and Roman military practices in discussions of more contemporary military problems. He likens the impact of the Swedes on Tilly, Wallenstein and Piccolomini to Pyrrhus' effect on the Romans.[92] He compares Louis XIII's organization of armies to Republican Roman practice.[93] He comments on Louvois' administration in these terms: 'the development of this system will promptly prepare for the return of the same abuses which brought about the decline of Roman armies' (*Le développement de ce système préparera promptement le retour des mêmes abus qui ont amené la décadence des armées romaine*s).[94] For Carrion-Nisas, like the authors discussed above, there is a seamless dialogue between the ancient military past and more recent times from which his contemporaries could profit professionally.

Slowly, writers, while referencing ancient experience, responded to contemporary military realities. Friedrich Rüstow (1821–1878) was a Prussian officer of progressive views who was cashiered and imprisoned for his support of the 1848 revolution in Prussia. He acknowledged the differences between his time and the ancient past due to the nature of ancient weapons.[95] Rüstow's belief that the military experience of the Romans was still relevant is evident from his work on Julius Caesar. He claims that the aspects of armies of his day, regardless of changes, must have existed in the past.⁹⁶ Rüstow characterizes the Swiss as adopting 'the good old Roman tradition' (*die gute alte römische Tradition*).⁹⁷ In the United States of America, Judson published his work on Caesar's army. This work explicitly compared what was known of Caesar's army, relying heavily on German scholarship, to contemporary armies.⁹⁸ Nicolas de la Barre Duparcq (1819–1893) was a French military engineer who taught at the École Spéciale Militaire de Saint-Cyr. His work, *Éléments d'art et d'histoire militaires* of 1858, continues the tradition of putting ancient military experience at the service of contemporary needs. De la Barre Duparcq assumes that ancient military arrangements can be defined as *ordonnances*, i.e. prescriptions, regulations or sets of rules, consistent with French practice.⁹⁹ In a discussion of de Saxe's advice that Prussian military practices be adopted by the French, he notes that the Romans became powerful because they were willing to adopt the practices of others, echoing Polybius.¹⁰⁰ He claims that Roman infantry carried 45 kilograms of equipment, noting that Davout's troops in Russia in 1812 carried only 30.¹⁰¹ Duparcq's belief in the continuing value of the study of Roman military experience finds echoes in Frölich's study of Caesar's military knowledge. In updating Rüstow's work, he aims to inform both scholars and officers.¹⁰² Steinwender studied the marching order of Roman forces based on a manipular organization. His rationale was that the battles of Jena and Sedan showed how important it was that best practice was identified.¹⁰³ He aims to learn from history rather than use classical texts to influence his military practice.

A most influential theorist who looked at the Romans for professional guidance was Ardant du Picq, a French officer who died as a result of wounds in 1870. Following his death, his writings became very influential in France and further afield including in the United States of America.¹⁰⁴ Du Picq, based on his studies of ancient warfare and modern, believed that the *élement moral*, i.e. 'morale', was, across the many changes in the military science and technology and different leaders, the primary consideration for success.¹⁰⁵ Du Picq's work, commonly known in English as *Battle Studies,* devotes a chapter to the manner in which the Romans' knowledge of men led to their military success. His work is more inspired by what he saw as Roman practice than guided by its tactical application. Du Picq's focus on morale echos von Berenhorst's critique of the Frederickian system.

The Emergence of Ancient History and the Roman Army

The academic discipline of Ancient History, as it is known today, is a relatively recent development. The ancient past was studied from the Middle Ages by reading ancient texts. Printing made these much more readily available in the Renaissance but even in the English universities of Cambridge and Oxford, the study was confined to commentaries on ancient texts, not history as we know it today.[106] This approach certainly served history well by the attention paid to the collection of ancient texts but what has been termed historical Pyrrhonism took a severely sceptical view of the trustworthiness of ancient texts in the seventeenth and eighteenth centuries.[107] Underlying these critical practices was the view that both antiquarians and historians had to establish the truthfulness of ancient texts. Today's idea of history as a conversation between past and present was established only slowly but, as we have seen, practical military officers had long been engaged in this project.[108] Even while Roman military experience was being studied by those with an interest in the profession of arms, the modern discipline of Ancient History started to emerge. Barthold Niebuhr (1776–1831), for Roman history, and August Böckh (1785–1867) for Greek, had already begun to establish the canons of the modern study of the history of the ancient world. Interest in the classical world was reflected in the scholarly output from Mommsen, Marquardt and their collaborators, which included very detailed treatments.[109] Adolph Becker (1796–1846), who taught archaeology at the University of Leipzig, began to write a five-volume work on Roman antiquities, appearing in the 1840s, completed after his death by Mommsen and Marquardt. In France, Fustel de Coulanges produced his work on the ancient city. Marquardt and Mommsen's volumes continuing Decker's *Handbuch der römischen Alterthümer* appeared in the second half of the nineteenth century.

It, and the huge scholarly effort required to produce it, reflected an interest in the ancient world for itself, while appreciating the similarities and differences between the that world and the modern.[110] We have already seen that military professional interest in the Roman army after the Napoleonic Wars demonstrated a more historical focus than tactical. In 1876 Marquardt released volume 5, part 2 with a section devoted to *Militärwesen*: the approach was entirely historical, not military professional.[111]

While works dedicated to the scholarly rather than professional military study of the Romans' military institutions began appearing, the interaction between soldiers and historians was already fluid. Scholarly attention had been given to the Romans' military institutions in the eighteenth century. Charles Le Beau had presented to the Académie des Inscriptions et Belles-Lettres twenty-six memoirs on historical matters between 1752 and 1777.[112] Rocquancourt, who we discussed above, was at pains to present the ancient military past to his students rather than rely simply on a reading of the ancient texts. In this, he reflects a growing historical interest in the ancient world, for its own sake. He may be the earliest historian to state, mistakenly, that Marius introduced cohorts into the Roman army.[113] He identified Marius as something of a change agent for Roman military arrangements, introducing what Rocquancourt referred to as the *ordre de Marius*.[114] Rocquancourt also saw a 'decline of the Roman military' (*décadence de la milice romaine*), echoing Vegetius' view of Roman history. In 1859, Rheinhard produced a book on Greek and Roman military antiquities for schools; it consisted almost entirely of illustrations. Clovis Lamarre's 1863 work on *la milice romaine* was addressed to a general public.[115] Lindenschmit's brief 1882 work was devoted completely to a historical description of the Romans' military institutions, reflecting the patriotic interests of Imperial Germany.[116] His work includes extensive illustrations, often of soldiers, from monuments. The enormous efforts of scholars in Europe included copying the illustrations on monuments like Trajan's column.[117]

There remained a belief that both military and historical interests in the Roman army were valid, although contemporary views on military affairs had a significant impact on how the Romans were understood. Karl Löhr was a junior officer in the Bavarian Gendamerie Korps who wrote for young military personnel, to address what he believed was a difficulty in accessing the military knowledge of the Greeks and Romans.[118] Löhr's work is one of the earliest which was devoted entirely to the ancient world rather than engaging with contemporary problems with assistance from ancient writers. It was common for the educated to have a knowledge of classical languages but a lack of access inspired Lipowsky to make Vegetius available to young Germans in 1827. He believed that a knowledge of Vegetius would be of value for both a civil and a military education.[119] De la Chauvelays' work published in 1884 is also devoted to the Roman army and engaged with de Folard's and Guischardt's earlier views on its operation. The work included an introduction from Léopold Davout, Duc d'Auerstadt, a French general who commended the work to both historians and soldiers.[120] Only three years later, Harkness' historically focused work on the Romans' 'military system' appeared to support his edition of Caesar's commentaries. Like Lindenschmit, he provides readers with detailed illustrations.[121]

Scholarly interest in military arrangements of the Romans became an increasingly academic pursuit in the early twentieth Century. A particular distinguishing feature of these works was language which implied that Rome's armies could be described in terms of the armies with which these scholars were familiar. This led to the use of terms like 'the Roman army' and similar terms in German and French works: *Römische Heer* or *Armée romaine*. Alfred von Domaszewski (1856–1927) was an eminent Austro-Hungarian philologist and historian, perhaps best known for his *Geschichte der römischen Kaiser* of 1909. He was a collaborator with Mommsen and Otto Hirschfeld and an important contributor to the *Corpus Inscriptionum Latinarum (CIL)*. He produced the *Die Rangordnung des römischen Heeres* in 1908. His title, which translates as *The Hierarchy of the Roman Army*, assumed a single Roman army and he wrote that Augustus had laid down the basis for that institution.[122] He, perhaps with the multi-ethnic Austro-Hungarian army in mind, saw Roman *auxilia* as mercenaries and that the *Offizierkorps* (officer corps) was needed to exercise authority over the men.[123] In Germany, Ritterling and Kubitschek's entries on the Roman army in the *Real-Encyclopädie der classischen Altertumswissenschaft* appeared.[124] They provide the foundation upon which almost all later works on the Romans' military institutions are based, directly or indirectly. The acquisition by European states of lands in other parts of the world is reflected in other work on the Romans' armies. Cagnat updated his 1892 work, *l'Armée romaine d'Afrique et l'Occupation militaire de l'Afrique sous les Empereurs* in 1913.[125] His research corresponded to the French possessions in North Africa. Cheesman's 1914 study of the *auxilia* is the earliest attempt at a comprehensive treatment of a component of Rome's forces. For Cheesman, the *auxilia* were transparently analogous to the sepoy regiments of the British Indian Army.[126] Haverfield's obituary for Cheesman, who died in 1915 at Chunuk Bair after landing at Suvla Bay, noted that he consulted military men of his own time and examined the Roman army in relation to other armies.[127] The first work in English to build on the work of German scholars on the Roman legions was published by Parker in 1928. Parker is explicit in his debt to German scholars.[128] Parker's countrymen continued to study the Roman army with a notable interest in Roman settlements in England, particularly those associated with Hadrian's Wall. Birley's brief note in the *Antiquaries Journal* of 1931 demonstrated the industrious scholarship of British scholars researching 'their' Roman Army.[129] The possibly unconscious assumption that the armies of the Romans constituted an institutional entity like twentieth-century armies remains even now in almost all works which address the Romans' military institutions.

Kriegswesen

During the nineteenth century, an old domain of knowledge, military science, *art militaire*, was enlivened by the development of what contemporaries referred to as *Kriegskunst* or *Kriegswesen*. It developed and can be seen as an extension of the interest in Rome's armies for professional purposes, using history as a discipline. Most military writers came to see the Romans' military experience as more a point of reference than as examples which could be directly applied. Most, like Du Picq, used the Romans' experience to inform debates on contemporary military problems as had European soldiers for centuries. The difference was that soldiers like Du Picq no longer looked for tactical inspiration but for much more general guidance on the conduct of war. This new view aimed to take a rigorous, scholarly approach to the conduct of military operations to improve military efficiency and effectiveness. In so doing, it willingly engaged with the past for the benefit of the present. As has already been mentioned, Rüstow, publishing in 1856, aimed to educate his largely civilian audience to understand warfare.[130] Rüstow's study of military knowledge and leadership of Julius Caesar explicitly compared Caesar's armies with contemporary armies.[131] Von Berneck's 1867 work on *Kriegskunst* was designed for officers, acknowledging the difficulties in accessing historical sources.[132] Von Hardegg's work, published a year later, is detailed in its treatment of Rome's armies and includes illustrations of troops, camps, weapons and equipment and siege weapons.[133]

Following the Franco-Prussian War, work by a Prussian Major General, Friedrich August Paris, was translated into French. The translators noted Paris' belief in the value of military history for officers but, while acknowledging the value of Greek and Roman history for general principles, Paris saw no practical value in Rome's armies for military science in his time.[134] On the other hand, Jähns, a scholar, in his *Handbuch einer Geschichte des Kriegswesens von der Urzeit bis zur Renaissance*, noted the view of the Chief of the German General Staff, von Moltke, criticizing those whose enthusiasm for practical military studies causes them to see the study of military history to be worthless for officers.[135] Wilms's study of the Battle of Cannae (216 BCE), a matter of keen interest to von Schlieffen, a later Chief of the German General Staff, aimed to resolve conflicting accounts of the battle amongst scholars.[136] Wilms believed that philology was not sufficient to address a military problem which needed professional understanding.[137]

A key figure in the use of history for military science is Hans Delbrück (see Figure 6).[138] His work on the art of war has been widely read, having been translated into English and Russian.[139] Delbrück came into conflict with the military history section of the German General Staff over his belief that he could speak authoritatively on military history without being a serving officer; he was a reservist, commissioned during the Franco-Prussian War.[140] Delbrück also had difficulty persuading his colleagues at the Berlin University that military history was not a mere *Fachwissenschaft*, technical knowledge, but *Geistwissenschaft*, part of the domain of the humanities.[141] There was a tendency among German scholars, many of whom were heavily influenced by Prussian traditions, to read ancient texts as technical knowledge.[142] Delbrück's contention that military affairs could not be well understood unless within a broader political and cultural context alienated both military and political opinion in Germany.[143] Delbrück's historical approach was thoroughly Rankean, materialistic and critical. Delbrück did not so much describe the 'Roman army' as explain Roman military success as a function of the Romans' socio-political institutions, criticizing ancient historians like Mommsen who were not sufficiently critical of the ancient sources.[144] Delbrück's practice of *Sachkritik*, combining source criticism with technical understanding of military operations, led him to a more-critical view of the accounts of ancient authors.[145] Delbrück however did not escape the danger of assumptions which reflected his own time. He discussed the Roman manipular organization, as described by Polybius, in terms of the drill with which he was familiar.[146] He also assumed, consistent with recent European military experience, that what he termed *Cohortentatik*, cohort tactics, required a change from a *Burgerheer*, citizen army, to a *Berufsheer*, professional army.[147] He explained the success of the Romans against Hannibal in terms of an *Offizierkorps*, as in the German army of his day.[148] He was explicit about his assumption: in his chapter on *die Centurionen*, Delbrück compared Roman centurions to French and British eighteenth-century practice as distinct from what he terms *heutigen Armeen*, modern armies, which had a distinction between the officer corps and men (*zwischen Offizierkorps und Mannschaft*).[149]

Veith's study of Caesar's campaigns owed something to Delbrück, believing that the 'the actual principles of war' (*eigentlichen Grundsätze der Kriegskunst*) remain the same.[150] Veith criticized the historian Mommsen because military details had been inadequately covered in his treatment of Caesar's campaigns.[151] In 1928 Veith published, with Kromayer, *Heerwesen und Kriegsführung der Griechen und Römer*. This work separated political and cultural issues from military considerations.[152] Kromayer and Veith were at pains to distinguish their work from what they termed *antiquarischen Darstellungen* (antiquarian representations) from *Heerwesen und Kriegsführung* (knowledge of armies and warfare).[153] Like Delbrück, Kromayer and Veith approached their task in an academic manner, acknowledging the work of other scholars. There remains a further example of the utilization of the Romans' military experience for their German contemporaries. The Carthaginian victory over the Romans at Cannae became, for the German empire, the solution to its great fear of a war on two fronts. For von Schlieffen, the Chief of the German General Staff, 1891–1906, a Cannae-style battle of annihilation could protect the Reich.[154] The General Staff published Schlieffen's collected essays regarding Cannae in 1925.[155] An English translation was provided to US officers in 1931.[156] While the Schlieffen Plan had failed against France in 1914, the strategy met with considerable success in the Battle of Tannenberg. The commander of the German 1 Corps, von François published a popular work on the battle in 1926, entitled *Tannenberg; Das Cannae des Weltkrieges in Wort und Bild*[157], i.e. 'Tannenberg, the Cannae of the World War in word and picture' (see Figure 8).

The work of scholars like Delbrück, Kromayer and Veith on Rome's armies has not been significantly advanced in military science, as opposed to history. The technological advances of the First World War and later have rendered the study of Roman military arrangements increasingly irrelevant to the concerns of modern military leaders. Hanson exemplifies this with his justified criticism of Delbrück and his German colleagues as relying too much on their contemporary German experience.[158] However, there are still some works which have used the Romans' experience to advance modern concerns. Luttwak's work on 'grand strategy' in the

Roman Empire attempted to advance a view of strategy through an interpretation of Roman military arrangements, particularly in the Eastern Mediterranean.[159] This work has had a mixed reception and generated a sometimes-bitter debate. Luttwak's approach, although strongly supported by some, has been seen as an attempt to apply a modern construction to the Romans' actions without understanding that the Romans had a different view of the world.[160] Nevertheless, the debate shows that the influence of the Romans' military experience remains on modern military thinking.

Conclusion

Since Roman times, writing about Rome's armies has been a source of a range of ideas from moral instruction to tactics and strategic policy. For European soldiers, classical texts about Rome's armies have been a source of inspiration and guidance. These texts were considered to be a ready source of advice until the appearance of the mass armies of the Revolutionary and Napoleonic Wars, even at the cost of disregarding the impact of gunpowder on warfare. At the time when Roman military experience was seen to be less and less relevant, the study of ancient history, as we know it today, appeared as an academic pursuit. Its more military manifestation in Germany was in *Kriegswesen* while in other nations, it became entirely civilian. Today, with only occasional attempts to reference Roman military experience to military or strategic problems, the study of Rome's armies is the domain of scholars, both professional and amateur, and of enthusiasts. For all, it is timely that the modern discovery of the 'Roman Army' is reviewed to reconsider the appropriateness of the terms and language with which we have become familiar.

Chapter 2

The Roman Army?

Army: a Simple Word Loaded with Meaning

It will not have escaped readers' attention that this work refers to Rome's armies, not 'the Roman Army'; an explanation is due. 'Army' is a simple word but it has a number of different meanings in modern English. If someone refers to the British Army in relation to the United Kingdom in 2023, images of soldiers, both in camouflage uniforms and in parade uniforms for the monarch's review or a coronation are likely to come to mind. Further, viewing one's nation's soldiers brings memories for those who have served and who still serve. Those memories are beyond reminiscence; they refer to what are often valued national institutions in which millions have served in the last 150 years. In these institutions, men and women have spent parts of their lives and in some cases lost their lives, separate from the rest of society, often performing duties which for the rest of the nation are extremely unpleasant. 'Army' can refer to a closed society with its own code of behaviour which has generated the sub-discipline of military sociology.[1] Many scholars and popular usage refer to the 'Roman Army' as a shorthand to distinguish it from the armies of other ancient nations or people. There may be little harm in that, provided that other implications of the usage are avoided; on the other hand, avoiding the usage avoids confusion.

The first problematic implication to be avoided refers to chronology. The Roman state, *senatus populusque Romanus*, i.e., 'the senate and the Roman people', existed across a vast period. Even if a description of the Roman army is limited just to a period from the First Punic War to the death of the emperor Honorius in 423 CE, that is still a span of almost 700 years. If we accept the Romans' dating of the state's foundation date of 753 BCE, we have a span of over 1,100 years *ab urbe condita* (i.e., 'from the foundation of the city'). By comparison, consider applying the same time span to a modern state, the United Kingdom or France for example, let alone the USA or Russia. Arguably neither the UK nor France has such a longevity. Both states experienced political changes including civil wars across that time period and so did the Romans. Both have existed for only 700 years, but it is highly unlikely that anyone with a knowledge of the military history of either nation would apply the terms British or French Army to the armed forces of either state across that whole time period. Using the term 'Roman Army' without a chronological qualification runs the risk of suggesting that it is meaningful across over a millennium and ignores the many changes which took place.

The second implication is to assume that the Romans viewed their military forces in much the same way that citizens today view their national armies: as distinctive social institutions which separate members from others in society. Indeed, meetings of former ex-servicemen and women can be, somewhat unfairly, caricatured as assemblies of different versions of Plautus' boastful *miles gloriosus*, Pyrgopolynices. However, what modern veterans share is a mutual recognition of a common experience of life different to most in the community. This is often tacitly acknowledged by others who accept that former and serving members of the armed forces have made a special contribution to the nation through their service. But this is not how the Romans saw soldiers or other functions in their lives. The Romans' world was divided sharply into *res privata*, and *res publica*, i.e. 'private things' and 'public things', or matters managed by the family and things common to all families. In a modern state, governmental and constitutional arrangements and people's professions separate religious, judicial, educational, military and political functions into separate domains. Within the Roman concept of *res publica*, with very few exceptions, there were none of the separate religious, judicial, political or military domains which characterize modern states. Pyrgopolynices was a butt for humour because of his preoccupation with just one aspect of the life of a Roman citizen, quite apart from his comic conceits. To be a Roman male citizen was, *ipso facto*, to be a soldier among other things.

In common modern parlance, 'the army' refers to a national army. The term 'Roman army' can unintentionally be taken to imply that the Romans had a national army in the quite-recent modern sense. Prior to the nineteenth century, European nations did not exist as much more than the possessions of rulers, and armies were rulers' armies. This is not to say that people in the nations or ethnic groups had no concept of their distinctiveness; the English, who may have developed a national consciousness in the sixteenth century, certainly understood themselves to be very different to the Scots, and vice versa.[2] But nations have not been ruled by their people until very recent times. For the ruling classes of nations prior to the French Revolution, the identity of the state was generally inseparable from that of the king. The English, having executed their king, still could not conceive of themselves without a king following Cromwell's death. Even religious belief could be regulated by the ruler. In 1555 at Augsburg, wars between Catholics and Protestants ceased, at least temporarily, with an agreement which can be summarized by *cuius regio, eius religio*, that is 'whose kingdom, their religion'. This formulation explicitly defined a ruler's territory to be ruled by the ruler's conscience; the views of others were not relevant because they were subjects. The formulation was confirmed later by the Peace of Westphalia in the different treaties which finally concluded the European wars of religion in the seventeenth century. Nations were what their rulers made them. The French Revolution may be taken as a turning point in Europeans' conception of their identity.[3] The confessional identities which defined Western Europeans since the Wars of Religion changed into identities defined by language and a sense of nation defined by literature, supported by rising literacy and access to printed material.[4] The corollary of this development is that, prior to the French Revolution, armies represented not their nations but their rulers. The implication of this is that there was, for example, no 'French' army prior to the Revolution; it was the army of *le Roi de la France* (the king of France), as opposed to the *l'armée Française* (the French army) that it would be in the nineteenth century under Louis Phillipe, who was *le Roi de les Français*, (the king of the French). *La*

France is a geographical expression of territory; *les Français* refers to a national group. Prior to the French Revolution and even for some time later, states were defined dynastically and so were the armies which fought for the states' rulers. As we shall see, authors prior to the nineteenth century did not refer to 'the Roman Army' in a nationalist sense. Is it reasonable to assume that the Romans' armies were national armies, in the sense of European armies since the nineteenth century?

Armies were, of course, more than just the soldiers.[5] The administrative apparatus required to maintain armies in peace and war is substantial. European states in the decade prior to the First World War were notable for the influence of their armies and the accompanying administration on their societies.[6] The armies were permanent, with their own administration and extensive records. Even earlier, Napoleon's armies were sustained by a vast administration which evolved to sustain his empire's armies.[7] Henri Clarke, Napoleon's Minister of War, built a framework for the administration of French armies, the influence of which survived into the twentieth century.[8] Other nations developed similar infrastructures to support the huge armies required to defeat Napoleon. From Napoleon's time, with the exception of Britain, the great (European) powers required young men to spend some years in compulsory peace time military service. For the French, military service became a means of creating national identity.[9] The German tradition was quite different. The success of the Prussians in 1870 resulted in the elevation of the Prussian king to German emperor and the imposition of the Prussian military code throughout the empire.[10] The influence of the army became central to German national life, integrated deeply into social and political institutions (see Figure 9).[11] Germany was by no means unique although emulated.[12] The victory of the Prussians, and their allies in what became Germany, over the French in 1870 showed the superiority of the short-term conscript army based on universal service over the French long service model which favoured the wealthy.[13] Internal conflicts in French society were reflected in the difficulties introducing a new conscription system.[14] The British saw military service only for local defence and relied upon their navy and maintained this view throughout the nineteenth century. The US took a similar view and relied upon poorly regarded professional soldiers and a large navy.[15]

The influence of nations' armies pervaded European nations prior to 1914 and the war that followed only emphasized the identification between citizens and their national armies. In the late nineteenth century, soldiers were looked upon with pride as evidence of modernity. In nations with conscription, military service became a rite of passage. In Britain, lacking conscription in the nineteenth century, the achievements of empire endowed the army and the navy with an aura lit by a similar national pride. Modern national armies are still seen in a very similar way and exist in an institutional sense. Beyond being national institutions controlled by governments, they are organs of the state controlling a particular aspect of national life. Given the chronological length of the history of the Roman state, from city state to the Tetrarchy, it would be quite inaccurate to assume that the Romans had a national army similar to today's national armies. Further, the much-simpler organization of the Romans' *res publica* until the death of the Emperor Augustus in 14 CE similarly makes comparisons with modern armies problematic.[16] The key issue to be discerned is whether the Romans saw their military arrangements as a social institution, like modern armies. With this question in mind, how did the Romans describe their military forces? Did they use language which suggests the existence of a Roman Army as a distinct national institution?

What did the Romans call their Armies?

There are a number of Latin words which can refer to 'army'; these four are most common: *exercitus, agmen, milites* and *militia*. In general, *exercitus* is a term for a collection of soldiers and it does not denote an institution.[17] Vegetius's use of *exercitus* in this sense is clear when he explains their composition.[18] Velleius, by distinguishing between the armies of the Roman people and the armies of the Italian enemies, also shows that *exercitus* refers to assemblages of troops rather than an institution.[19] Ancient writers used *exercitus* to refer to armies commanded by particular persons or distinguish armies from fleets.[20] There are some Vespasianic coins with the inscription: CONSEN/SVS EXERCIT/VS, i.e. 'agreement of the army', referring to Vespasian's single army before he was emperor.[21] Coins referring to plural armies again show no institutional sense of the term.[22] *Agmen*, another word interpreted in English as 'army' but without any sense of institution, has a narrower range of meaning than *exercitus*. It refers generally to the army moving.[23] It could refer specifically to the army's baggage or column of march.[24] The most common word for soldier in Latin is *miles*. Its plural form, *milites*, can be read as army in the sense of a collective noun but, again, it does not occur in the sense of the army as an institution. Like *exercitus*, it can refer to soldiers associated with a particular commander.[25] Finally, we come to *militia*. The word refers primarily to military service.[26] The word's import is carried by the dichotomous phrase *domi militiaeque*, i.e. 'at home and in military service'.[27] Juvenal's Sixteenth Satire is devoted to the benefits of military service for soldiers. The term Juvenal uses is *militia*: '*Quis numerare queat felicis praemia, Galli, militiae*', i.e. 'military service: who is able to count the benefits of fortunate military service, Gallus?'[28] Tertullian also uses *militia Caesaris* in the sense of military service for the Roman emperor and contrasts this with *militia Christi*, soldiers of Christ.[29] The Brigetio Table, dating from early in the fourth century CE, refers to *militia* in the sense of personal military service.[30] Ammianus uses *militia* to refer to armies composed of infantry.[31] Vegetius, in defining the subject matter of his second book, distinguishes *exercitus* and *militia* when he writes: '*Secundus liber ueteris militiae continet morem ad quem pedestris institui possit exercitus*' ('the second book describes custom of the old army according to which an infantry army may be formed').[32] The term *vetus militia* refers to 'old military service', as in Juvenal above, as opposed to Vegetius' times. By the Late Empire, well beyond the period of this study, *militia* had come to mean any state service,.[33] In that sense with military service becoming state service, Rome's armies may have become the state, but as numerous revolts and civil wars show, not in a national sense. If the Romans had a sense of their soldiers as members of a national institution in the modern sense, it is not to be found in the way they described them.

A particular feature of modern armies is that they are permanent, that is to say they exist in both peace and war. Even here, until Augustus, the Romans raised armies for particular commands, and lacked a 'standing army'.[34] There is an often-ignored account in Cassius Dio which should remind modern readers that Roman armies prior to Augustus existed for particular campaigns and were then disbanded. Dio invents a debate between Maecenas and Agrippa in 29 BCE. In the course of this debate, as a narrative device, Dio has Maecenas make a number of significant recommendations concerning the army and its financing. Among these is that Augustus should maintain στρατιῶται ἀθανάτοι (*strationtai athanatoi*), i.e. 'undying soldiers', armies composed of citizens and allies.[35] This force, Maecenas argues, should train continually. Dio tells his readers that Augustus decided to take Maecenas' advice, but he did not act at once because of the magnitude of the changes. Dio places the debate in 29 BCE, at a point when he thought that a critical change took place. It is very much a hindsight view because, as Dio states, in as much as it was implemented, it was done gradually from 29 BCE, not as a complete policy at a single time.[36] Dio's choice of language to describe this army shows that he wished his readers to understand that the proposal for, or the practice of having 'undying soldiers', was one which dated from Augustus' time. Dio and his readers in the ancient world would have known that the Romans had, prior to the time of Augustus, only raised troops for a particular command. At the end of that command, the troops would be disbanded or handed on to the next commander but were eventually disbanded when the command was terminated, i.e. they were 'dying' soldiers and armies. Dio is not alone among ancient authors in believing that 'something' happened to the armies in Augustus' time. At least one other author who wrote after Dio noted changes which he believed to be most significant in the history of the armies. Herodian tells his readers that Italians used to fight, but that Augustus took away their arms and instead set up camps where soldiers served for fixed rates of pay.[37] Herodian is highlighting the major features of the new

Augustan 'undying army', not the processes by which it came about, so his observation should not be taken, any more than Dio's vignette, to indicate a single moment of policy creation. The 'undying soldiers' also provide a context in which to understand Josephus' comments on the Roman army for his Greek-reading audience in the first century CE. He stresses that the Romans are always ready to fight, because they have an army continually in existence and do not have to raise one.[38] Dio's 'dying armies' insightfully describes a key change which resulted from Augustus' principate because, as we will see in the last chapter, the armies of Augustus were never disbanded and the practices adopted in these armies became the norm in the armies of the emperors who followed Augustus.[39]

Finally, compared to most modern states, Roman history is somewhat remarkable for its civil wars and revolts. While it may not be difficult to explain why a Sulla, Caesar or imperial pretender would seek to advance their political interests militarily, why would the soldiers follow them? For a modern state, such behaviour would be treasonous and contrary to the values of modern national armies committed to the defence of the nation, but not so for Roman soldiers. The simplest way to explain this behaviour is that Roman soldiers had no sense of corporate or institutional identity as members of 'the Roman army'. These soldiers saw themselves as Sulla's, or Caesar's or Vespasian's soldiers.

The term 'Roman Army' is meaningful if defined chronologically, if it does not imply the existence of a social institution as a national army, if it acknowledges that it served a particular person and to distinguish it from another nation's army, as the Romans did. Other uses are too imprecise and run the risk of unintentionally representing Rome's armies as modern national armies. To avoid confusion, I refer to 'Rome's armies'.

Chapter 3

Changes in Rome's Armies, the Problem with 'Reform' and an Alternative

Introduction

In the preceding chapters, we have traced the evolution of the study of the Romans' military arrangements and considered the problem inherent in describing their armies in the same terms as modern armies – that is to say, the difficulties in assuming that the Romans' armies can be discussed in institutional terms, as modern armies are. However, even if it is accepted that Rome's armies were quite different from modern ones, it still needs to be explained why and how they changed between the traditional foundation of the city (*ab urbe condita*) and the death of Augustus. In this chapter, we will examine the specific issue of the nature and explanation of change in Rome's armies. In modern accounts of the Roman army, change is often described as 'reform'. This is true from treatments of Servius Tullius to discussions of Augustus's activities related to the armies. Characteristically, the reforms are agglomerations of changes which are attributed to a reformer. This remains true even when the evidence is acknowledged to be weak. Scholars have struggled with this, but reform has not been abandoned in relation to changes in the Roman army.

As historians, we try to explain the past by constructing a narrative of Roman history, a story if you will. The different stories often depend upon larger, more encompassing stories, providing contexts which are often referred to as metanarratives or grand narratives. There is a current metanarrative of Roman history of successive reforms, political and military, then failures which end with the fall of Rome in the West and the continued existence of an Eastern remnant called the Byzantine Empire. Reform is commonly used to describe historical change within this metanarrative, not just in relation to the Roman army but also in more general works on Roman history for well over a century.[1] The term appears in the general histories which have defined the grand narrative of Roman history. It also occurs in monographs, journal articles and occasional pieces. It is so common that the modern approach to explaining change in the Roman army might be termed the 'reform paradigm' or 'reform model'. Reform as an historical concept in relation to the Romans' armies is characterized by a number of features:

1. Reform is defined as such within a time sequence of cyclic crises and responses.
2. Reform occurs within an institutional framework.
3. Reform occurs under a leader, the reformer, who orchestrates the changes.
4. Reform is achieved by specific measures, the effects of which are assumed to be policy-driven and predictable.
5. Reforms are designed for improvement of institutions but the effects can be broader.
6. Reform is often seen, consciously or unconsciously, as teleological, i.e., the reform is part of the explanation for progress in the metanarrative.

Reform assumes a particular view of historical events, even though its use has changed over time. The central question is the degree to which it is appropriate to use the term to describe change in Rome's armies. We will examine the import of the term 'reform' and its use in modern works. As we shall see, the Romans do not describe changes as reforms. Further, there is no evidence that the reforms in relation to the armies, as described in modern works, occurred. None of this denies that change occurred, but it does leave open the issue of explaining why changes occurred. In later chapters, we will examine particular claims of 'reform' associated with individuals like C. Marius and Augustus.

The Place of Reform in Roman History

Hooke's multi-volume history of Rome, which was initially published in the mid-eighteenth century refers to both 'reformation' and 'reform' but in the sense of correction of abuses.[2] The word 'reform' appears in Gibbon's late-eighteenth-century work on Roman history in the sense of reform of the state or of religion.[3] The term also appears in general histories of Rome early in the nineteenth century.[4] By the middle of the nineteenth century, reform is used by Niebuhr's editor Schmitz in the English translation of Niebuhr's work and in Smith's classical dictionary.[5] At much the same time, Thomas Arnold referred to political change as reform.[6] Liddell's Roman history of 1855 also used reform to describe change and so does the Englishman Merivale's text, published in the United States during their civil war.[7] Towards the end of the nineteenth century, reform continued to be used to describe change.[8] In the early twentieth century, reform remained a common way to describe changes.[9] In the period prior to and during the Second World War, reform was still used to describe change. In particular, Cary's influential English account of Roman history uses the term extensively, in thirty pages.[10] For example, it refers to the following reforms down to the time of Tiberius Gracchus:

- Reform of the Roman Army about 450 BCE;[11]
- Further Reforms of the Roman Army – manipular;[12]
- Reform in the *Comitia Centuriata* early in the second century BCE;[13]
- Reforms in the Judicial System in the second century BCE;[14]
- Neglect of Army Reforms after the Punic Wars;[15]
- Scipio Aemilianus opposes reforms, and[16]
- Tiberius Gracchus' land reforms.[17]

Cary is not unique in the use of reform to describe change.[18] The reason authors like Cary use this common term is because it reflects an accepted pattern in history, referring to both political and military developments. This use of reform to describe what are seen as major change points in

Roman history is not as common in recent works, possibly because it is assumed. Even if the most recent edition of the *Cambridge Ancient History* does not use reform to describe change points in Roman history in the same way as Cary, the term is still ubiquitous in the series to characterize administrative, financial, military and political change. Reform appears on twenty-seven pages in Volume 7 Part 2.[19] Volume 8 includes reform on ten pages.[20] Reform can be found on twenty-seven pages of Volume 9.[21] On eight pages of Volume 11, reform is to be found.[22] In Volume 12 the term is found on twenty-eight pages.[23] Reform appears on just four pages in Volume 13.[24]

Reform has been used extensively in modern works to refer to changes in Rome's armies down to 14 CE. While reform has been used less frequently in recent modern works to describe political change, it continues to be applied to the army, though here too the use of the term is decreasing. In the period covered by this study, the reforms to which reference is most commonly made are the Servian/Tullian, Camillan, Marian and Augustan. Not all scholars refer to all these reforms, but it remains that works on Rome's armies in recent decades still make reference to most of them. Forsythe refers twice to reforms assigned to Servius Tullius.[25] Keppie uses the term to refer to the change from the use of hoplite phalanxes to the use of a looser formation.[26] Carniart refers to Marius' reforms and even describes the changes to Roman cavalry in Polybius as reforms.[27] Goldsworthy refers to 'Marius' reform'.[28] The 1998 edition of Webster's work still refers to 'the reforms of Marius'.[29] Alston refers to 'Augustan reform'.[30] In Erdkamp's edited work on the Roman army published in 2007, Gilliver entitles an entire chapter: 'The Augustan Reform and the Structure of the Imperial Army'.[31] Farnum refers to Augustus' 'military reforms'.[32] Sometimes reforms are described as evolutionary, reinforcing the progressive nature of the changes.[33] Reference to reform is not restricted to the anglophone world.[34]

Reform as a Historical Concept and Rome's Armies

How appropriate is the word 'reform' and its associated concepts for describing changes in Rome's armies to the death of Augustus? There are at least two issues here: did the Romans conduct progressive changes and is there any evidence that the Romans instituted such changes? To situate our discussion, consider the manner in which reform is used in public discourse today to characterize change. For many political and other leaders in different walks of life today, reform is a term used to characterize a change. In particular, economic and political innovations, which it is claimed ought to be made, are called reforms.[35] How is it that this small word has come to carry these connotations? Of what value is this word in describing change in Rome's armies?

Reform has more than one meaning depending upon one's concept of history. In Latin prior to the Christian era, there is no word whose meaning is captured by 'reform', in its sense of change that is innovative and positive, as defined at the beginning of this chapter in relation to the Roman armies. There are a number of words in Latin which capture some senses of reform: *correctio* or *emendatio*, both of which really mean 'to free from mistakes, improve, straighten out'. The expression '*meliorem/melius facere/efficere*' occurs, as in Seneca quoted below. There is no progressive sense implying a teleology in these terms. For positive actions to correct a bad situation, a word for Romans would more likely be *reddo* (restore*)*, *redeo* (go back, return) or *restituo* (put in its former place).[36] *Corrigere* (set right), and *reficere* (restore), are common. *Reformare*, the word from which the modern term 'reform' is derived, first appears in Ovid's *Metamorphoses*, referring to a transformation back to a previous state and elsewhere to a rejuvenation of an old man for one day.[37] This is the Latin version of the Greek μεταμόρφωσις, metamorphosis.[38] Valerius Maximus uses *reformare* for the restoration of Athens by Themistocles after the Persian invasion.[39] Seneca refers once to '*reformatio morum*':

> Lucilius, I am accustomed to do this: from all enquiry, even now if philosophy is taken away for a long time, I try to discover and to

bring about usefully. What of yours, which we drew out just now, is more removed from a change in customs (*a reformatione morum*)? How are Platonic ideas to make me better? What from yours may I derive what may suppress my desires?[40]

The sense is very much of change of an individual's character and perhaps by extension, people generally, relative to an existing ideal. Seneca assumes that in the past Plato provides the benchmark for customs. Language, like *restituta* and these uses of *reformare*, refer to change as looking back, if anywhere, but it is certainly not looking forward. The same can be said of contemporary views of the problems of the Roman Empire in the third century CE and of coins of all periods. Some scholars have argued that in the third century CE some contemporaries believed that a solution to Rome's problems lay with energetic emperors who were prepared to restore traditional values. This was not seen as a 'reform' in the sense of progress.[41] In the military sphere, changes in weapons and equipment are discussed in ancient texts. Livy tells his readers that the Romans changed their shields from round to rectangular, began to serve for pay and substituted the maniple for a battle line, i.e., *phalanx*, like the Macedonians.[42] Apart from the error in the nature of the *phalanx*, Livy does not present these changes in a progressive sense. The works of Xenophon and Ammianus on the Persians in different periods, Asclepiodotus and Onosander on Macedonian tactics and organization, Polybius' treatment of the Romans' armies and Vegetius' advocacy for a reversion to past practices, have no implications of a sense of progress in military affairs.[43]

Reform is not part of ancient discourse except sometimes for the Christians looking to the second coming of Christ, an implicitly teleological view of the world.[44] Augustine gave *reformare* its teleological cast.[45] Until the fourth century, *reformatio* was not even a strong feature of the Christian vocabulary. Tertullian could suggest *quantum reformavit saeculum istud*, i.e., 'how much he restored the age', looking back.[46] *Reformatio* referred to here was about a return to tradition in very traditional Roman terms: *ad pristinam gloriam reformare*, that is 'to restore to ancient glory'.[47] Man's change to conform to the likeness of God is a central tenet of the concept of reform in Christian thought.[48] A decisive change came when Augustine insisted that it was possible for Christians to improve beyond the innocence of the Garden of Eden, i.e. the Christian development of an old idea: *reformare ad melius*, or 'restore for the better'.[49] *Reformatio* became fallen men's effort to improve their world and it did not look back to the past. It became a penitential concept; men reformed their lives.[50] As Ladner says, 'Contrary to all vitalistic renewal ideas, the idea of reform implies the conscious pursuit of ends'.[51] This teleological view of Christian salvation history informed individuals' view of their lives but it has little impact on how communities saw their world.[52] During the Middle Ages, there were within the Church, calls for reform but these were in the Roman sense of restoration to standards set in the past. Following the Renaissance, an intellectual debate arose in Europe which pitched those committed to a view of the world founded in interpretations of the Greco-Roman classics against those who promoted a criticism of those ideas. In 1617, Cawdrey defined reform in terms of change, censure or correction.[53] The modern discourse of reform, appearing in the late seventeenth century but becoming significantly influential during the eighteenth-century Enlightenment, initially focused on correction of abuses but later moved to a future-orientated desire for improvement in personal and social relations. In 1828, the *Webster's Dictionary* defined reform as:

> To change from worse to better; to amend; to correct; to restore to a former good state, or to bring from a bad to a good state; as, to reform a profligate man; to reform corrupt manners or morals.[54]

This was a similar, if expanded, version of Cowdrey's definition but by 1913 a new element had been added, reflecting a change in meaning and usage compared to 1828. The definition in the 1913 edition of *Webster's Dictionary* begins 'To put into a new and improved form or condition'. This reformulation introduced the idea that the 'improved form or condition' might be 'new', i.e. an innovation, and reveals a progressive sense not referred to in the past. In 1959, reform was defined in these terms, without explicit reference to 'new', in a scholarly work on the concept in Church history prior to the Enlightenment: 'the term and idea of reform are applied to the renewal and intended improvement of many things, more often however of social entities and of institutions than of individuals'.[55] The author goes further in describing reform in more recent times:

> Since the Enlightenment, however, and especially since its alliance with the biological idea of evolution in the nineteenth century, the idea of progress has acquired connotations of continuity, irresistibility, and all-inclusiveness which are lacking in the concept of reform.[56]

This future orientation reflected the replacement of the humanists' role models with progress.[57] This made reform inherently linked to progress as human improvement broadly.

Reform promoted the development of the structures and machinery of the modern, Western state during the nineteenth century.[58] It became the alternative to revolution with its commitment to the improvement of the European state, exemplified in the massive legislative changes undertaken by successive British governments from 1832 to 1918 known as the Reform Acts.[59] In British military affairs, Edward Cardwell, Secretary of State for War on Gladstone's Liberal-Whig ministry (1868–1874), introduced a series of changes which were commonly referred to as the Cardwell Army Reforms by contemporaries.[60] Reform as a rationale for change continues to be used in the twenty-first century, but it now carries more the sense of 'a change for the good', with a lessening of confidence in progress in the developed world.

For all the benefits and the usefulness of the discourse of reform in the recent and modern worlds, it must be accepted that there are no Greek or Latin texts, before Augustine, which advocate for reform as a manifestation of progress. Quite the opposite, there is no general sense of progress which would justify innovation.[61] In fact, for Cicero, *res novae*, i.e. 'new things', were tantamount to revolution.[62] *Tabulae novae* ('blank slates') were feared because they abolished the debts on which the wealthy depended.[63] In ancient texts, the past is often seen as a more glorious time which the present could aspire to emulate, as so many ancient authors tell us, with only Thucydides as a contrary voice.[64] The Greeks did recognize a limited kind of progress in the spheres of philosophy, sciences, arts and constitutions but not in moral or political activity.[65] The Romans did not accept the idea of progress, implicit in the modern use of reform, as we do, except perhaps in technology.[66]

The Etruscans believed that their nation would have eight 'ages'. The last age would be one of decadence, a common enough view among Roman writers. Plato's theory of forms looks not to progressive improvement but to a preferred state in the past. Similarly, in the *Laws*, Plato imagines an ideal community seeking simplicity, not innovations.[67] The myth of Prometheus in *Protagoras* presents human abilities and tools as having a divine origin.[68] The Roman Lucretius shows a sense of decadence but he also shows a kind of progress in human insight, at least his own.[69] Some arguments used to oppose the destruction of Carthage demanded by Cato included the idea that Carthage's existence would prevent Rome sliding into decadence.[70] Polybius' theory of anacyclosis is a well-known example of a process that ends or begins in decline. Cicero refers to the decline of the state.[71] Both Sallust and Cicero use the phrase *inclinata rei publicae*, 'decline of the state'. Sallust sees a biological metaphor: *omnia orta intereunt*, 'everything which begins dies'. Under the empire, Seneca the Elder describes Roman history as a life cycle beginning with the monarchy and ending in old age with the principate.[72] Emperor Marcus Aurelius believed that a sensible man of forty years of age would have seen all that will and could be.[73] Even militarily, Diodorus records that Philip II acknowledged Homer as the source of his phalanx's pikes, while Frontinus saw no need for improvement in siege equipment because of the exhaustion of human invention.[74] This was not a world that saw change for improvement, the assumption implicit in progress, as desirable or even possible.[75]

Reform as means of progress is not used by ancient authors to describe change in the Roman army or of anything else for that matter. This, of itself, does not mean that the use of reform is inappropriate for modern use. There are a great many concepts we use that are not used in the ancient world. The use of the term with its progressive connotations may have seemed appropriate for the nineteenth and much of the twentieth century. At that time the history of the Roman world was seen as teleological, and progress was seen as beneficial. In regard to military history, the modern interest in technological progress has too often fallen into technological determinism, quite inappropriate for Rome's armies.[76] However, with a term as loaded with meaning as reform, particularly meanings which do not comply with ancient world views, to describe change in the Roman world in such terms requires us to either claim a greater understanding of their world than the ancients had and/or to claim a critical explanatory power in the concept which cannot be otherwise provided by alternatives. For historians, it is methodologically challenging if not anachronistic to explain the past in terms of the present even though history continues to be a conversation between the past and present. To describe something as a reform rather than a change implies that what happened was progressive to some end and beneficial by some criteria. It is by no means inappropriate for historians to make judgements about historical events, though judgements about the ancient past of ancient armies are probably of dubious value. On the other hand, to describe a change as simply change carries no implicit judgement of the change, its value or significance. It may be true that some historians have assumed that the discourse of reform is simply an alternative to change. However, such an assumption is at best careless and at worst shows a lack of insight and an awareness of the manner in which contemporary views of history are shaped by previous historical experience as well as our own.[77]

There were changes, often referred to as 'reforms' in Rome's armies between the foundation of the city and the death of Augustus. A reform characterized a change as part of a progressive solution to a problem. As such, reform implied a reason for change as a step in the march of progress. While the progressive element in reform may no longer apply today, it remains that the term carries the associated ideas about how change is implemented, i.e. policy-driven attempts by a central figure to make things better. While evidence of particular reforms will be considered in subsequent chapters, the abandonment of the reform paradigm is challenging for two reasons apart from a lack of evidence. If the progressivist element is abandoned, as it mostly is today, we lose the implicit explanation for historical change supplied by progress. Secondly, the abandonment of reform also leaves us with no way of explaining how change occurred. Any alternative to the characterization of change as reform demands not just a different mechanism for the initiation and execution of change, it also requires a different explanation of the underlying causes of change in the Romans' armies.

An Alternative Explanation for Change: Contingent Decision-Making.

We must return to the basic questions of what changes occurred and how they are to be explained. In the at least 800 years under examination, there appear to have been comparatively minor changes in equipment and organization but these do not seem to have altered significantly the manner in which the Romans fought.[78] If reform is discarded as a means of characterizing change, the changes in the Romans' armies that did occur still need to be explained. It needs to be understood that reform, as a concept, is more than a way of describing change. It also characterizes change as progressive, even teleological. The abandonment of the reform paradigm, then, is a problem for historians for two reasons. If the progressivist element is abandoned, as it mostly is today, we lose the implicit explanation for historical change supplied by progress. Secondly, the abandonment of reform also leaves us with no way of explaining how change was carried out. Any alternative to the characterization of change as reform demands not just a different mechanism for the initiation and execution of change, it also requires a different explanation of the underlying causes of change in the Romans' armies. Evolutionary metaphors have an attraction because they, being progressive, provide an implicit explanation of historical change and may even suggest why change occurred. The problem with evolutionary explanations in history is that they are only metaphors. Evolution, as a theory, and its mechanism of natural selection operate at species level, not at the level of historical institutions unless dubious social Darwinist theory is applied. What we seek is something akin to Thucydides' truest, underlying, cause as well as a series of immediate causes.[79] However it is one thing to explain the causes of a single war in terms of underlying and immediate causes; it is quite a different thing to explain change over 800 years in this manner. The underlying forces, which sustain these changes, have to be pervasive and have also to incorporate the immediate causes of change. The solution, applying Occam's razor, lies in the nature of armies raised for limited times and defined purposes.

The truest cause of the change in Rome's armies were the contingent decisions made by Roman generals as a function of the scope of their authority. Most of the decisions are unknown to us and, even of those about which we do have knowledge, most had few long-term consequences. However, some of these responses to local circumstances became established practices and it is these that we see as changes. The accumulation of contingent decisions produced adaptations to the circumstances in which the armies operated. The immediate cause of change in Roman armies was the nature of Roman military leadership and its requirement for generals to operate in a broader set of administrative domains than modern military commanders. It has often been assumed that leadership and decision-making in the Roman world was very much like that of modern First World states in the nineteenth and twentieth centuries. In such states, generals relied upon centralized defence establishments to provide logistics, personnel, intelligence and training support for military operations. While Roman commanders were issued instructions from the Senate, or later, from the emperor, they could not rely on anything like the same scale of support provided to modern armies. The Roman state did not have the extensive bureaucracy and the accompanying communications which modern states enjoy. Roman generals operated within a different framework. Local arrangements addressed most logistic, personnel, training and intelligence functions. The degree to which Roman government was bureaucratic has been subject to different modern opinions. Modern treatments of the administration of the Roman state, particularly in the Imperial period, have focused on the existence and growth of a bureaucracy. There can be no denying that some sort of what in our terms is a 'bureaucracy' existed to provide administration for the empire. The issues of contention are the size, the scale and especially the mechanisms and functions of the bureaucracy.[80]

The first three centuries of the Empire have been described as 'government without bureaucracy'.[81] There is no reason to believe that their description does not apply to the Republican period too. We know of nothing that would suggest the existence of centralized bureaucratic military records before 14 CE. This leaves us with a position in the period covered by this study that there was no substantial modern style bureaucratic structure on which Roman generals could rely to support their military operations. The focus of Roman military bureaucratic procedure, to the extent to which it existed, must have been the local unit. Unlike modern service records, soldiers did not have unique numbers, files do not appear to have been kept alphabetically or in some other standard manner. It is likely that most documents were not meant to be used by anyone other than the clerks who maintained them.

A very important support provided for modern generals by their countries is in the provision of information and personnel to aid operational decision-making. This is all the more so in the data-rich environments in which modern commanders operate. Some have assumed, for example, that the Romans in the Imperial period would have had a centralized military intelligence agency, as modern states do, but it remains that there is no evidence that such an institution existed. Roman generals and provincial governors were responsible for their own intelligence.[82] In these circumstances, generals still had to make decisions but did so contingently, i.e. when they needed to due to the force of circumstances and in the context of limited information. Again, there is no reason to believe that generals were any better supported prior the death of Augustus. Even if the Senate issued instructions, the implementation was made at local level, by individual commanders. If this is true, then changes may have come about as responses to specific problems. It is of course possibly the case that Roman commanders often made decisions after consultation with the Senate in Rome. There is every indication that this occurred regularly, but the speed of communication in the ancient world would not have permitted Roman commanders to seek advice on operational issues which were either too trivial and/or too urgent to be dependent upon a reply from afar. Roman generals were not expected, nor were they often able, to defer to the Senate in anything like the degree to which generals in modern armies do to their governments. Instead, Roman generals were expected to make do with whatever resources were at hand. Military success in part depended upon a general's capacity to do this. This is a characteristic of all pre-modern armies to a degree. This is not, however, to be overstated. All generals in all times are required to respond to the realities they are confronting. The key distinction is that modern generals can rely on considerable support, though that can also translate as organizational inertia and interference. Pre-modern generals did not have this support, so they were required to rely on their own resources to a greater degree. Generals had to address both the normal administration of their armies and to adapt to the circumstances in which their armies operated. The contingent decision-making to manage day-to-day issues saw a degree of micro-management unknown in modern armies, while the contingent decisions required to adapt to circumstances produced changes of many kinds, at many different levels of importance, some of which were adopted as normal practice and so became permanent.

The contingent decision-making practised by Roman generals had its basis in the nature of their authority. Almost all Roman generals were magistrates and even those who seized power styled themselves as such. The basis of their authority was *imperium*, the authority to command troops in the field, conferred on magistrates by the *Comitia Curiata*. However, when examining the legal powers of Roman magistrates, we need to be mindful of the dangers of trying to apply modern legalistic assumptions to ancient practice. The Romans did not share modern constitutional approaches to the powers of magistrates. Customs of the ancestors, *mos maiorum*, was the most significant influence. There are a number of ways of describing the powers of Roman generals. Mommsen extrapolated the powers of *imperium* in the military sense from the statement of Cicero: ... *imperium, sine quo res militaris administrari, teneri exercitus, bellum geri non potest*, i.e. '*imperium* without which it is not possible for military matters to be administered, armies to be held, war to be waged'.[83] The paradigmatic military nature of the functions of *imperium* is to be found even in Virgil.[84] In the nineteenth century, Mommsen, who cast Roman government in constitutional terms, saw *imperium* in the Republican period as having a number of components:

- raising and building an army,
- appointing officers,
- war leadership,
- making agreements and treaties for allies and enemies,
- administration and financial management,
- battlefield command including authority over soldiers and civilians,
- the right to the title of imperator if successful,
- the right to a triumph,
- the right to bring an army back to Rome at the successful conclusion to a campaign.[85]

Some of these aspects involve the administration of the magistrate's *provincia*, i.e. province, and others are more specifically military. An equivalent of Mommsen's list of duties and rights is not found in the ancient world for the Republican period. What we do have are texts which describe the desirable qualities of a Roman general and a small number of texts on generalship itself.

The most detailed exposition on the ideal qualities of a Roman general comes from a political speech. Cicero's *Pro Lege Manilia* extols Pompey's virtues as a general.[86] Cicero in this speech provides a number of different lists of the qualities required of a general and from these we can deduce something of what qualities Roman generals were expected to have. Cicero uses *virtus* (plural: *virtutes*) both in a specific sense of qualities desirable in a general and in the general sense of the qualities the Romans expected in men. The four *virtutes* Cicero believed a general should have are *scientia rei militaris* (knowledge of military matters), *virtus* (manliness), *auctoritas* (authority) and *felicitas* (luck).[87] In a later set of qualities, Cicero lists: *belli scientia* (knowledge of war), *singularis virtus* (personal manliness), *clarissima auctoritas* (clear authority) and *egregia fortuna* (outstanding luck).[88] The last three qualities in both lists need not detain us for our current purpose, except to note in passing that they are personal and reflect upon the character of the commander. The first requirement in these lists, *scientia rei militaris*, points to what we might identify as professional knowledge.[89] A similar view is found in Livy when Aemilius Paullus replies to his critics that he did not object to advice from the wise and those with expertise in military affairs, *proprie rei militaris peritis*, i.e 'particularly experienced in military matters'.[90] What was the *scientia rei militaris* to which Cicero and Livy, in different terms, referred? Roman aristocrats around Cicero's time could learn *scientia rei militaris* in the field and most probably did. There are a number of texts which indicate how young aristocrats obtained such knowledge. Pliny the Younger, in the Imperial period, describes the practice of earlier times by which young Roman aristocrats learnt about leadership in general, but it is notable they are described as learning in a military situation. Pliny suggests that in earlier times, young aristocrats learnt to lead in the camps and in the field from a young age.[91] While perhaps an idealized view, during the Republican period aristocrats, might spend years in military service and so learn basic skills in this way.[92] Cicero, however, also expresses regret that young men even in his time no longer spend much time soldiering.[93] While Cicero did not condemn the use of books, he reminded his audience in another speech that Marius did not need books to learn about war.[94]

There are three types of ancient works which address *scientia rei militaris*. All seem to be directed at generals or potential generals, most show generals at work and all show the key role of the general in the micromanagement of his forces. All of these works, while including some older material, date from the period after Augustus' death but there is nothing to suggest that the views expressed would have been different in our period. The first group describes the organization, training and manoeuvring of armies. The works in this category include those of Aelian, Aeneas Tacticus, Arrian, Asclepiodotus, Cincius, Frontinus, Julius Africanus, Mauricius, Pseudo-Hyginus and Vegetius. Of these, Cincius' work is too fragmentary to be of any assistance.[95] The other works are informative in different ways. The focus of the works of Aelian, Aeneas Tacticus and Asclepiodotus and Arrian's Τέχνη Τακτική (Techne Taktika, Tactical Handbook but usually referred to as *Tactika*) is on Hellenistic tactics but Aelian's work and *Tactika* contain some Roman material including, in Arrian's case, details of Roman cavalry training. The presentation of these writers' material for a Roman audience shows that the actions of ancient generals were universally relevant.[96] Arrian's Ἔκταξις κατὰ Ἀλανῶν (*Ektaxis kata Alanos*, or *Tactics against the Alans*) describes in some detail his campaign against the Alans in the second century. Pseudo-Hyginus' *De Munitionibus Castrorum*. (*Concerning the Fortification of a Military Camp*) provides our most detailed ancient treatment of Roman encampment practices. There are other works which have not survived. Vegetius tells us that he based his work on the writings of Cato, Cornelius Celsus, Frontinus (probably in his lost *De Rei Militari*), Paternus and the *consititutiones* of Augustus, Trajan and Hadrian.[97] These works have not survived. Emperor Mauricius' Στρατηγικόν (*Strategikon*), while addressing circumstances well outside our period, is the most comprehensive surviving manual written by a Greek or Roman general.

The second group of second and third century CE works deal with advice to generals, as in the stratagems of Frontinus, Julius Africanus, Onasander and Polyaenus. These treatments stress the moral qualities, a reflection on the value of ideology or values, as much as, if not more than, the technical knowledge of a commander. This has surprised some modern scholars who have expected a more technical approach to military leadership.[98] What these ancient works show is not just the technical aspects of military activities, they also reflect the cultural values within which military knowledge was a domain of aristocratic prestige.

The third group of works contains the substantial accounts of Roman military operations in biographical and historical works. These accounts, based in whole or part on generals' *commentarii*, can supply additional information though we need to take care. Some of these works are by individuals who had first-hand experience of Roman generalship. These include the surviving accounts of military operations by Caesar and the author of *De Bello Hispaniensi* (*About the Spanish War*), Velleius Paterculus' treatment of the campaigns in which he participated, Josephus' account of Roman operations in Judaea and Ammianus' account of Julian's operations in the East in the fourth century CE. Caesar's works and those attributed to him may be the closest that we get to the content of generals' *commentarii*, even allowing for the self-advertisement implicit in the works. The nature however of such works is to further underscore the role of the general.

So much of the material was written in the Imperial period, outside the scope of this study, but there is nothing to suggest that the Romans prepared men for military command differently prior to the death of Augustus. Beginning with the first group of works, although referring to the Imperial period, Arrian's *Tactika* is a collection of material which stresses Greek military practice but includes Roman material. His *Tactics Against the Alans*, a brief account of a battle formation to be used against the Alans, may be unusual in its style but its content shows no inconsistency with what we know of contemporary Roman military practice. Though the extant text does not describe its purpose explicitly and lacks a dedication, it nevertheless appears to describe an actual commander's account of the deployment of his troops and the subsequent battle against cavalry raiders. As it stands, this work would suggest that a commander would attend to quite minor details in his decision-making to deal with tactical situations. Similarly, Pseudo-Hyginus' *De Munitionibus Castrorum* is a detailed treatment of Roman encampment practices. Its attention to technical detail is similar to what is found in modern military manuals but, unlike modern manuals, it shows no evidence that it was centrally produced by a training establishment.

As regards the second group, which provides advice for generals, the work of Onasander is typical. The work, which looks at the task of a Roman commander, is dedicated to the Roman consul of 49 CE, Q. Veranius, a man with a successful military career of his own.[99] This work, while in Greek and making no attempt to be original, does seek to offer a Roman commander genuine guidance. In a discussion of the qualities of a general, Onasander lists the military tasks to which a commander should attend and again the focus is on micro-management. Onasander assumes little military knowledge from his readers except perhaps weapon-handling skills. For example, he stresses that generals need not give all orders in person but, instead, use a chain of command.[100] Such advice is elementary to modern readers with any military experience, but the fact that he feels the need to offer it suggests that his audience may have had little military knowledge. Even under the Principate where we might expect that a general's work would be more circumscribed, Onasander expected a general to take an interest in and micro-manage so many aspects of his army's activities on his own initiatives. There are references in the sources to the *constitutiones* attributed to Augustus, Trajan and Hadrian, though none has survived. The fact that Vegetius refers to them suggests that he used them as sources but it is unclear to what extent. The references in Arrian to Hadrian's preferred cavalry drills and Hadrian's comments in the inscription at Lambaesis suggest that he laid down what was to be done and how.[101] Again we see a Roman general, in this case the emperor, concerning himself with micro-management. Vegetius' reference to these works does not indicate the degree to which they were comprehensive, though the descriptions in Arrian's cavalry exercises implies that Hadrian's instructions may have been quite detailed. There may have been equally detailed instructions for infantry units. Arrian refers to a work he wrote for the emperor on Roman infantry exercises which may have formed a basis for Hadrian's own instructions to his armies or perhaps have been inspired by them.[102] Whatever was the nature of Augustus' and Trajan's *constitutiones*, Hadrian felt the need to draft something different. Dio's comments on the continuance of Hadrian's practices in his day are in the context of their acceptance and use by the soldiers.[103] This tells us something of the function *constitutiones* served. Like 'standard operating procedures' today, they provided frameworks in which soldiers and generals operated.

Vegetius' work is one of only two surviving Latin works whose title, *Epitoma Rei Militaris* ('*Summary of Military Matters*'), suggests that it aimed to address *scientia rei militaris* in general. The other work is the anonymous *De Rebus Bellicis* ('*About Warlike Things*'). Frontinus' lost *De Re Militari* is a possible source for Vegetius. The title of Vegetius' work claims that it is a summary of military matters. The content of the work is truly comprehensive and detailed, regardless of any modern claims of archaisms. Vegetius covers everything from choice of recruits, their training and equipment and organization through to battle formations and siege warfare. As an indication of what constituted *scientia rei militaris*, it is invaluable. The anonymous *De Rebus Bellicis* reveals both the limited nature of the resources available to the unknown emperors to whom the work is addressed and the assumption that an emperor would take an interest in details of equipment as readily as he would be concerned with taxation or laws across the empire. After an initial five chapters on expenditure, corruption and finance, and before two final chapters on frontier defence and the legal system, the author discusses in detail a series of machines and devices, which, he believes, will aid the defence of the empire. The fact that this work and Vegetius' cover such a range of matters shows that the authors believed that an emperor would himself have responsibility for such a range of issues. *Scientia rei militaris* on this basis is a broad field. There is no indication that the emperors had anyone on whom to rely for such technical knowledge apart from their own experience and their close advisors (*proximi*), though these could be very experienced in military affairs.[104] Consequently, works such as those by Vegetius and *De Rebus Bellicis* find their place by providing collected advice. Pseudo-Hyginus' *De Munitionibus Castrorum* has a much narrower focus but provides very detailed advice on laying out a Roman camp and is probably the most technical of all extant Roman military works. Some surviving ancient manuals are similar to those on which modern soldiers depend. Works like those of Vegetius, Arrian, pseudo-Hyginus and the *constitutiones* seem quite similar to the kinds of manuals in use in Europe over the past 150 years. Like modern manuals, these ancient handbooks are designed to support training and military operations but unlike their modern counterparts, they were not developed by centralized training establishments.

There is a whole body of ancient works aimed at commanders which have few equivalents in modern military training literature. These are the books of stratagems by Frontinus and Polyaenus. Frontinus, who lived a little later in the first century CE than Onasander, produced a work, *Stategemata* (*Stratagems*), which is more comprehensive in what it tries to address than Onasander's.[105] Both works address the practical issues which a general might face in the field. Frontinus' work differs from Onasander's in using examples of stratagems from the past under the headings which provide the structure for his work, whereas Onasander's treatment is discursive. Polyaenus's work also contains stratagems but with less organization than Frontinus and a focus on Greek examples, even including the gods. Stratagems are the epitome of contingent decision-making. They address specific, one-off situations. In modern warfare, such things are almost unknown except for deception plans to hide intentions, but they have nothing like the significance accorded to stratagems in ancient military literature. A reading of Frontinus and Polyaenus shows the vast ambit of this ancient path to military success. Included in the lists of stratagems are not just deceptions to be worked on friend and foe, but there are also changes to what we refer to as organization, equipment and conditions of service. The chapter headings in Book One of Frontinus' *Strategemata* are:

- About hiding plans,
- About discovering the enemy's plans,
- About determining the situation of the war,
- About leading an army through enemy infested places,
- About getting out of difficult situations,

- About treachery on the march,
- How to conceal what we lack, do without or find an alternative,
- About distracting the enemy,
- About suppressing mutiny,
- How to obstruct an untimely request for battle,
- How to enthuse an army for battle,
- About dispelling the fear which soldiers gain from unfavourable omens.[106]

Julius Africanus' Κεστοί (*Kestoi*, approximately meaning 'compilation') contains similar stratagems and associated information of use to a general.[107] Polyaenus' work is organized differently. It records stories about and stratagems used by Greek and Roman generals. His treatment of Scipio Africanus is typical. On the one hand we read how he deceived an opposing army into standing-to for hours and only attacked when his enemy was tired and hungry; on the other hand we also discover how he imposed camp discipline by expelling prostitutes and laying down the kinds of food the troops could eat.[108] The stratagems in both works demonstrate the expectation that a Roman general would attend to quite small details in the micro-management of his soldiers, using contingent decision-making.

We must now consider the third and largest group of works: biographies and histories. Among works by authors who had first-hand experience of Roman military operations, there are numerous examples of Roman commanders using contingent decision-making to meet needs. The changes they made consisted of everything from small changes in equipment to larger changes in organization; what we should note is that the generals were quite prepared to make such changes. Caesar's works provide the most-substantial, most-detailed accounts of the operations of Roman armies in the Republican period. Caesar made many adaptations to his circumstances by using contingent decision-making. These decisions extended to the organization of his troops and even to the supply of equipment to address particular problems. At one point during his civil war, Caesar formed a temporary unit of *expediti* from *evocati*, archers and slingers.[109] This temporary formation may not have been unique. There are numerous references to *expediti* in the corpus of Caesar's works but their composition is rarely described. A notable example of improvization is the mounting of *legionarii* as cavalry to protect Caesar from Ariovistus.[110] Later Caesar placed 400 e*xpediti antesignani* (literally 'in front of the standards', that is ahead of the main battle line) with his cavalry as a reinforcement to enable him to defeat Pompey's cavalry.[111] Of course these temporary formations could have led in different circumstances to more-enduring arrangements. This is the manner in which temporary contingent decisions could lead to permanent change. In Africa when fighting the Pompeian forces there, Caesar trained his troops to deal with the numerous skirmishers.[112] Caesar's troops were not alone in adapting to local circumstances. He notes that the Pompeian troops who had long served in Spain had adopted Spanish styles of fighting which stressed skirmishing.[113] In Caesar's works, he is depicted making and observing tactical changes, but not only these. In Africa, Caesar even saw to the manufacture of weapons, and missiles in particular.[114]

The accounts of Roman military operations in the manuals, collections of stratagems, biographies, panegyrics and histories provide many details of contingent decision-making by Roman generals. However, we must be cautious. These works reflect the social and political prejudices of their authors' times. The fact that they were probably often dependent upon generals' *commentarii*, works whose aims were at least partially to highlight the achievements of their subjects, must cause us to wonder about what others in the armies did. Yet the fact remains that we know little of the activities of those others in relation to change in the army. While we cannot rely on the argument from silence, we can at least say that generals were contingently adaptive, even if unknown others were too, and that because of the generals' role in the military hierarchy, they were most likely to be the key agents of change.

Implications for our understanding of the Roman army

The Romans' armies changed from earliest times to the death of Augustus in 14 CE. The most likely mechanism for specific changes was the role of

Roman generals. Roman generals were expected to do what was necessary to win battles and to govern their provinces. Their powers came from the *imperium* which they enjoyed. With these powers they were expected to deal with the normal activities of administering and conducting the affairs of their armies. This required a degree of micromanagement rare in modern armies. Whether they were administering their province or commanding their armed forces, Roman commanders adapted to their circumstances by making whatever changes were necessary. They had few people or institutional resources on which to rely in these circumstances. Some of the changes they made became common practice more broadly; most did not. None of these changes were reforms. If an army continued to exist for a relatively long period with little contact with other Roman armies, the result could be a different Roman army in form, like the Pompeian forces in Spain. It could be argued that the kinds of changes I have described, arising from the actions of Roman generals, could and have been made by generals throughout history. This may be correct, but we would not describe such small changes as reforms. This is the point of the examples above. The instances of micromanagement described in the stratagems are small and contingent. Over time, many small, contingent decisions produced changes.

Chapter 4

The Myth of Professionalism in Rome's Armies before 14 CE[1]

Introduction

We have considered the difficulties which can arise from the use of modern terms to describe ancient entities. The concept of professionalism is used commonly in twentieth- and twenty-first-century scholarship with reference to the armies of the Romans. The armies and their soldiers are seen by scholars to have become increasingly professional from the mid-second century BCE until Marius took the step, at the end of the century, of enrolling the *capite censi,* the lowest census class, which, it is claimed, made soldiers more dependent upon military service as a livelihood, promoting greater professionalism. Augustus is viewed as having taken the final step of making the armies fully professional. While the language of professionalism is common, there are a range of opinions on the degree to which the soldiers in Rome's armies were professional prior to the time of Augustus, Rome's first emperor who dominated Roman politics until his death in 14 CE.[2] Scholarly opinions are virtually unanimous that the 'Roman Army' was professional after Augustus, and increasingly so.[3] This almost paradigmatic view of Roman history invites the question as to whether 'professionalism' and its cognates are the appropriate terms to apply to Rome's armies from 200 BCE to the Battle of Actium in 31 BCE. It is not a merely semantic quibble. The application of modern terms to describe ancient realities risks misunderstanding aspects of the ancient world through the unwarranted assumption of correspondences between the contemporary world and the past.

Professionalism: a Modern Concept

Professionalism in relation to institutions is an idea of the last 150 years. 'Professional' did not appear in common English usage until the nineteenth century and 'professionalism' not before 1850. Professionals initially were seen as mercenaries.[4] In its general current use, professionalism implies a social obligation that produces both group solidarity and limits to membership based on demonstrable expertise.[5] In relation to language, professionalism and its cognates imply a moral orientation to the community of an occupational group, emphasizing a degree of altruism.[6] Professionalism thus reflected the appearance of professions, which some view as a necessary component for the development of government institutions in Europe from the sixteenth century.[7] For Western democracies, the concept has been seen as an aspect of the legitimisation of power and the consolidation of the power of the state, normalizing the role of the citizen in the economic system.[8] 'Professionalism' is, and has been, applied to many areas of human activity since it was first coined. Military professionalism is a particular application mostly within Western European states, but it has been a recent development. As in other professions, it distinguishes amateurs from those with a more-dedicated commitment to technical knowledge and expertise.[9] The broader development of increasing specialization in Europe in the eighteenth and nineteenth centuries in technology, particularly industrial, and the division of labour led to a significant growth in the complexity of military and naval operations.[10] The growth in the size of armies required specialists in logistics to coordinate the movement of bodies of troops in space and time.[11] This, along with the changes in armaments in the nineteenth century, led to expectations of professionalism in the armed forces in many countries.[12]

The increase in the complexity of European warfare in the wake of the Napoleonic Wars of the nineteenth century, with technical improvements in weapons and the use of railways, required soldiers to develop greater expertise. Dedicated training for officers had previously been rare. Most officers learnt 'on the job'. This is exemplified by the historian Edward

Gibbon, who in his memoirs recalled that his service as a captain in the Hampshire militia during the Seven Years War assisted his later work as a historian of the Roman Empire.[13] Gibbon's training, unlike that of officers in almost any modern army, was entirely regimentally based and his qualification for his rank was a function of his social class rather than his knowledge and skill at arms. In the nineteenth century, many states responded to the need for better-trained officers with the creation of training institutions with set curricula to prepare professional officers.[14] This was particularly true in relation to the challenges of mass mobilization, for instance, where the Prussian General Staff demonstrated particular skill in preparing for and conducting the war with Austria in 1866. In some Western nations with a tradition of parliamentary democracy, a tension developed between professional armed service officers and democratically elected or appointed officials.[15] The treatment of the French Jewish artillery officer, Alfred Dreyfus, by conservative military superiors caused a scandal in France early in the twentieth Century and led to direct intervention by the democratically elected president and closer public scrutiny of the army.[16] In Britain in 1914, a revolt by aristocratic cavalry officers at the Curragh Camp delayed the provision of home rule for Ireland, leading directly to the Easter Rising in 1916.[17]

In all states that accepted professionalism in their armed forces, an implicit social contract was struck by which civilians promised the military autonomy in return for civilian control.[18] According to Sarkesian and Connor, writing about the United States military of the twenty-first century:

> In an official capacity ... it is the officer corps that is involved with any number of civilian groups and institutions affecting civil-military relations. The officer corps sets the reference points in the military community and shapes the environment in which military families live. It is the officer corps that is accountable and responsible for military effectiveness and institutional character. Thus, the military profession and the officer corps become virtually synonymous."[19]

The power of the discourse is such that military history has been cast as the progressive control of military activity by professionals.[20]

Professionalism and Rome's Armies, from Earliest Times to the Death of Augustus

When scholars use the term 'professional' in relation to Rome's armies from earliest time to 31 BCE, what do they mean and how justified is the use of the term? No ancient source refers to the armies of the Romans as 'professional' or uses similar terms. Study of the ancient world has been a feature of European learning for millennia, and interest in the Romans' military arrangements has long been common among military officers, but the study of history as we know it today evolved during the eighteenth century, as we have seen.[21] The adoption of the discourse of professionalism, when speaking of Roman armies, is entirely the work of scholars since the middle of the nineteenth century, at much the same time that scholarly specialization in what was termed the 'Roman army' also developed.

Montesquieu, in the eighteenth century, wrote that *'agriculture et la guerre étoient les deux seules professions des citoyens romains'* ('agriculture and war were the two sole professions of Roman citizens'), showing no knowledge that the Romans had professional soldiers or armies.[22] Gibbon, writing at the end of the eighteenth century, refers to the 'military profession', 'profession of arms' or 'of soldiers'; or of the law; or even of gaming, but he does not refer to professional soldiers.[23] A distinction needs to be appreciated here between 'profession', as a calling or employment, and 'professionalism' as a discourse, discussed above. Ferguson, writing at much the same time, refers to 'professional soldiers' twice in reference to Augustus' time, but in a manner that is uncomplimentary to both citizens and soldiers, reflecting the earlier implication of professionals as mercenaries.[24] Hooke, one of the earliest English historians to publish a history of Rome in six volumes in the mid-eighteenth century and republished into the nineteenth century, found no need to use 'profession' or 'professional' at all.[25] Goldsmith, whose work was published ten years later, uses 'profession' only three times and 'professional' not at all.[26] Niebuhr, who worked mostly in Prussia, does not refer to professionalism in either the Romans' armies or their soldiers.[27] Michelet's work appeared in France in the 1830s. He refers to *métier*, i.e. 'job', 'trade', or 'profession' in the sense of profession of arms, once in his first volume; in his second volume, he uses *métier* in this sense once too.[28] Michelet's compatriot, Todière, released a short work on Roman history that does not use either 'profession' or *métier* in the sense of profession of arms, let alone 'professional'.[29] Across the channel, Thomas Arnold uses 'professional' twice in his first volume.[30] In his second volume, in a discussion of ancient constitutional arrangements, he refers to the profession of arms but emphasizes that 'the business of a soldier was no isolated profession, but mixed up essentially with the ordinary life of every citizen', in effect denying the sense of separateness implicit in the modern concept of 'professional'.[31] Mommsen, in the first volume of *Römische Geschichte*, does use *Beruf*, usually translated as 'career' or 'profession', but sometimes simply as 'job', but not *Berufssoldat*, i.e. professional soldier.[32] In the second volume, *Beruf* and *Berufung*, calling or vocation, appear, but neither is used in the sense of the profession of arms.[33] Mommsen does indicate the distinction between *Berufssoldaten*, career soldier, and *Bürgersoldaten*, citizen soldier, but he does not refer to Roman soldiers as professionals.[34] He does not refer to *Berufssoldaten* in his work on the provinces either.[35] The 1871 English translation of Mommsen's work does use the terms 'profession' and 'professional', but does not apply them to either the armies or the soldiers.[36]

Consistent references to professionalism in Roman Republican armies date from the late nineteenth century and the appearance of specialized studies of Roman military activities. Dodge's 1894 work on Caesar's army presents the change to professional soldiers as 'less reliable material entered the ranks' compared to 'the burgess-soldier of the simple republic', and 'Caesar's legionary was no longer a citizen-soldier, as in the Punic wars; he was a professional, or a mercenary'.[37] The 1894 English translation of Mommsen's history refers to the development of a 'military class' and 'military service became gradually a profession'.³⁸ Leighton, in 1889, refers to warfare as 'a regular profession' and the establishment of a 'standing army'.³⁹ Myers refers to professional soldiers replacing citizen soldiers in relation to the work of the censors of 443, whom he links with the provision of *stipendia*, military pay.⁴⁰ Delbrück's 1920 work, *Geschichte der Kriegskunst im Rahmen der politischen Geschichte*, devotes an entire chapter to *Berufs Armee: Cohorten Taktik*, or 'Professional Army: Cohort Tactics'.⁴¹ Tenney Frank, in his work of 1923, also refers to 'professional soldiers' but indicates that the Romans were 'afraid that if professional soldiers were kept militarism might endanger democracy', perhaps reflecting his view of the former German empire, just after the Great War.⁴² Kubitschek saw a link between the use of cohorts and a professional army.⁴³ Other scholars had also begun to find 'professional' Roman soldiers and 'professional' Roman armies. In 1928, Kromayer and Veith described Scipio Africanus's army in Spain as a *Berufsheer*, a professional army.⁴⁴ They also describe Carthaginian soldiers as *Berufssoldaten* and the Roman army of the Imperial period is termed a *Berufsheer*.⁴⁵

Since the 1930s, the view that Roman soldiers and the Roman army became "professional" appears to have been accepted with few dissenters. Parker's influential work, entitled *The Roman Legions*, references Delbrück to state that 'the institution of a professional army (*Berufsarmee*) was officially recognized'.⁴⁶ In volume IX of the *Cambridge Ancient History*, published in 1932 and covering Roman history from 133 to 44 BCE, editors Cook, Adcock, and Charlesworth, outlining the key features of the period in the preface, state that: 'These [defence problems] led to the creation of a formidable Fourth Estate in an army professional in recruiting and sentiment'.⁴⁷ Last, in the same work, refers to service in the army after the Jugurthine War as a 'profession'.⁴⁸ The 'Marian Army Reform' is presented as providing the Romans with soldiers 'who were something like professionals'.⁴⁹ These men are represented in these terms: 'When men make the army a profession, they cut themselves off from civil life', demonstrating an acceptance of the sense of professionalism which was discussed earlier.⁵⁰ Last identifies Augustus as the provider of the solution of the problems the authors see with professional soldiers.⁵¹

Cary, in his *History of Rome down to the Reign of Constantine*, initially published in 1935, sees Marius, at the end of the second century BCE, making 'reforms' that include:

> In throwing open the legions to proletarians on terms of voluntary enlistment, the training of his recruits up to the standard of regular soldiers, Marius took the decisive step in converting the Roman army from a conscript militia into a standing force of professional warriors.[52]

Scullard, whose work on the early history of Rome was first published in 1935, also notes in his 1951 edition a transition from 'citizen militia' to 'professional army', which he sees beginning when Roman soldiers are first paid for their service at the time of the capture of Veii in 396 BCE.[53] Syme, on the other hand, in his 1939 work *The Roman Revolution*, believes that the Romans 'were at least preserved from the dreary calamities that so often attend upon the theoretical study of the military art or on a prolonged and deadening course of professional training'.[54] In 1940 Adcock split the difference, writing that, in relation to Rome's armies, 'at least from the Second Punic War onwards, there were men who made soldiering their profession and supplied the centurions who were the backbone of the legions'. Adcock contended that these centurions displayed 'professional competence'; however, he also wrote that the armies 'of the middle Republic remained largely unprofessional' but they became 'more wholly professional in character' after Marius.[55] The use of 'professional' and 'professionalism' in relation to the armies of the Romans continued after 1945. Gabba, writing shortly after the Second World War, described the Roman army after Marius as professional.[56] Harmand agrees, arguing that Marius was responsible for a fundamental change:

> the year 107 saw the old armies of the type based on the census give place to a new army recruited from proletarians, this put in motion the replacement of the old system of levies for each campaign with permanent professionals.[57]

Brunt's work on Italian manpower uses 'professional' but criticizes Gabba's view that the former soldiers of Lucius Sulla, whom he settled as colonists, were professional soldiers.[58] His concerns are with the accuracy of Gabba's statement, however, not with the use of the term 'professional' in reference to the soldiers. Smith's work of 1958 sees the Romans under Augustus as having 'a professional army' but he contends that professional soldiers existed in the second century BCE, well before Augustus.[59] Smith sees a dichotomy between 'professional quality' and what could be provided 'on a militia basis'.[60] Brunt, in his 1971 discussion of Livy's Ligustinus, a Roman citizen whom he presents as a model of civic virtue, finds professional soldiers prior to Marius.[61] Brunt describes Augustus' forces as a 'professional army'.[62] Garlan devotes an entire chapter to 'Roman Professionalism', emphasizing 'proletarianization' and Marian 'reform'.[63] Keppie, in 1984, sees a 'growth of professionalism' beginning in the second century BCE, whereby 'a core of near professionals' existed; but he argues that 'the Romans adopted professional attitudes to warfare long before the army had professional institutions'.[64] Keppie interprets the speech of Marius in Sallust's *Jurgurtha* as emphasizing 'his professionalism'.[65] Campbell, in a review of the military manuals available to Roman commanders, suggests that ancient interest in 'professional expertise' demonstrates a lack of appreciation of the importance of 'technical details', which he assumes to be important.[66] Webster's work on the army in the Imperial period sees the Republic ending because of professional armies. He describes Ligustinus as a 'full-time professional soldier'.[67] Patterson characterizes Livy's Ligustinus as having 'a particular enthusiasm and aptitude for military service who volunteered regularly for service to build a quasi-professional career with the legions'.[68] Patterson finds a 'gradual professionalisation' that was 'accelerated' by the opportunities offered by the extraordinary commands of Pompey and Caesar.[69] Dawson, in his treatment of 'Western warfare', finds that wars in the Late Republic were fought by 'increasingly professionalized armies'.[70] For Webster, under Caesar 'the army had become a highly efficient and thoroughly professional body'.[71] Nevertheless, Webster holds Augustus responsible for the 'first fully professionalized standing army'.⁷² Goldsworthy sees a citizen militia being replaced by 'a professional army'.⁷³ He does see the professionalism as 'apparent' though.⁷⁴ Goldsworthy contends that the use of cohorts was 'made possible by the professionalisation of the army'.⁷⁵ Eventually, for Goldsworthy, 'the citizen militia of the Republic changed into a professional army of long-service soldiers recruited from the marginal elements of society'.⁷⁶

Feugère, writing in the twenty-first century, believes that changes accorded to Marius produced *une armée professionelle*, a professional army.⁷⁷ Hildinger agrees with this.⁷⁸ Lynn writes that 'ca. 100 BC, the demands of foreign wars became so great that the Roman army ceased to be a citizen militia and became a professional mercenary force'.⁷⁹ Brizzi, in 2004, sees the origins of the professional army in changes by Marius that accepted men '*prêts à faire du service militaire une profession*', that is 'ready to make military service a profession'.⁸⁰ He and Erdkamp both believe that Augustus professionalized the army, as does Rosenstein.⁸¹ Contrary to these views, both Rawlings and Hoyos argue that discontinuity of service makes a description of Roman soldiers as professionals inappropriate.⁸² Hoyos does find professionalism in the Carthaginian armies though.⁸³ Cagniart characterizes the period 146–30 BCE as 'the transformation of the Roman army from a militia-citizen to a professional army'.⁸⁴ For Cagniart, the 'true professional army was Caesar's making, the last stage of the transformation of the Roman Republican army'.⁸⁵ Lukas de Blois, noting Keppie's use of the term 'professional' to describe Roman attitudes to warfare, asks whether it is reasonable to apply this language to most Roman armies.⁸⁶ De Blois writes that while Marius' soldiers 'did not constitute a professional mercenary army … they became nearly as good as professionals by experiencing one military campaign after the other'.⁸⁷ This applied to the officers too, and to Caesar's army.⁸⁸ Alston, noting the view that Marius 'created a professionalized force' that became more politically dependent upon its generals, and that willingness to serve in the armies was caused by poverty, observes:

There was probably virtually no economic or sociological distinction between the soldiery of the mid-second century and those recruited by Marius. The Marian reforms did not mark a sea-change in the political nature of the army.[89]

Kate Gilliver, noting potential misconceptions, comments that 'a professional standing army such as Rome had must have been similar to later professional armies' and argues that Augustus established a professional army.[90] Cosme, in a discussion of veterans who enlisted repeatedly, notes that *'ces soldats pouvaient déjà être considérés comme des professionnels de fait'*, ie 'these soldiers can be already considered as professionals in reality'.[91] Cosme believes that Augustus *'annonça la création officialle d'une armée permanente et professionnelle'*, that is he 'heralded the official creation of a permanent, professional army'.[92] In this, Cosme, Southern and Dixon are in agreement.[93] Southern believes that Octavian (Augustus) 'effected the changes that transformed the Republic into the Empire and its former citizen militia into a permanent professional army'. Southern, however, seems to see experience as cognate with professionalism, presenting Ligustinus as an example of 'a class of semi-professional soldiers who enrolled in one army after another, campaigning in Spain, Greece, and the east' in the second century BCE, and suggesting that 'more professional officers emerged with extensive military experience'.[94] Southern, though, agrees with Campbell, 'that the Romans in the republican period did not see the need for "professional military specialists"'.[95] Milne believes that 'administrative reforms' made the 'Army of the Late Republic ... a professional force'.[96] In a discussion of unit cohesion in modern armies, Milne refers to the Roman army 'because it is potential connection to a family unit which forms a major point of difference between the early Roman armies, where service was temporary and seasonal, and the professional army of the Late Republic', implying that such service was neither temporary nor seasonal.[97] Roth believes that under Marius 'the Roman army was quickly transformed into a quasi-professional one' and that by Augustus's time, 'the Roman army had long been, in practice, professional'.[98] Matthew simply believes that Marius created 'Rome's First Professional Soldiers'.[99]

Should Roman Soldiers be Called Professional Prior to the Death of Augustus?

As we have seen, the concept of professionalism has been applied commonly to the armies of the Romans for over a century, but never in antiquity. Is this significant? While the term and its cognates have been used extensively in reference to Rome's armies, this use by scholars has hindered rather than added to an understanding of Rome's armies up to the Battle of Actium in 31 BCE. Professionalism, as a discourse, has a somewhat awkward fit to Roman armies prior to Augustus and is not accepted by all scholars.[100] But were there forces permanently in existence, i.e. a standing army in which professionals could serve?

Livy's account of the enlistment of soldiers for the campaign in Macedonia, which culminated in the Battle of Pydna (168 BCE), shows that there was no army in existence from which to draw men. The men were volunteers: '*multi uoluntate nomina dabant*' ('many gave names voluntarily'), and thirty-two who had previously served as *primi pili* (senior centurions) in legions were enlisted as *milites*, i.e. private soldiers, contrary to their expectations.[101] During an assembly convened by the Tribunes of Plebs, to whom some of the former *primi pili* had appealed, a former centurion, Spurius Ligustinus asked to speak. His account of his military service began with his entitlement to serve based on his property qualification as a land holder: '*pater mihi iugerum agri reliquit et paruom tugurium, in quo natus educatusque sum*', (my father left me an iugerum of land and a small hut in which I was born and raised), and his family circumstances. He then outlined the campaigns in which he served, but he was clear that each was a separate enlistment. He was made a soldier by enlistment ('*miles sum factus*' – 'I am made a soldier'), served, and was discharged ('*dimissi essemus*', 'we were discharged'). He then enlisted again ('*miles voluntarius*', 'volunteer soldier'), and a third time ('*tertio iterum uoluntarius miles factus*', ('for the third time made a voluntary soldier'), and even twice more ('*et deinceps bis*', 'and twice more').[102] Each of these enlistments were for separate armies, with ranks determined by the magistrate who held the command. Ligustinus stated that the *tribuni militum*, military tribunes, determined ranks. He accepted his obligation to serve, believing that his only exemption was that he has provided four replacements, his sons. Livy presents Ligustinus as a moral example. Neither Ligustinus nor any of the others whom he addresses were professional soldiers, as we would know them. Soldiering as directed by the magistrates was the soldiers' duty, not their profession, in the modern sense. Livy's depiction of the enlistment of men into the legions is consistent with Polybius's description.[103] Both men had witnessed the enlistment process, i.e. *delectus* ('selection'), in the first and second centuries BCE. We have seen earlier that Dio and Herodian noted that something changed in Augustus' time and armies were not disbanded but that does not necessarily make the soldiers professional.

The fact that Ligustinus and others were enlisted repeatedly shows that they were not full-time soldiers, a key characteristic of today's 'professional soldiers'. A reflection of the problems this causes for claims of professionalism is found in Miller's doctoral thesis on the professionalization of the army in the second century BCE, where military professionalism is defined as 'the continuous practice of the art of war by repeated enlistments into the legions'.[104] This definition is not consistent with how soldiers in other armies are seen historically. The conscript soldiers of the Napoleonic Wars in Europe had similar continuity of service but are not described as professional. Their predecessors in the European armies of the eighteenth century like that of Prussia, although enlisted for long periods, spent most of their time off duty and could even practise trades.[105] In the nineteenth century, private soldiers, who were mostly conscripts in all armies except the British and the US, were not considered to be professionals, although their officers were expected to be.[106]

Roman magistrates were the officers in Roman armies. Their functions were civil, military, and religious, with almost none of the specialization one would expect of professionals.[107] The implicit social contract, under which professional officers since the late nineteenth century have accepted civilian control in return for military autonomy, never existed in the Roman world because the categorical modern distinctions between military and civilian did not exist. The simple application of a modern term like 'professional' or 'professionalism' to ancient circumstances is anachronistic. This anachronism is sometimes explicit; Adcock attempted to explain the operation of Roman armies by specific comparison to English social classes and the English armies of the eighteenth century.[108] The difficulties arising from anachronism have been outlined by Cadiou recently in his trenchant critique of modern views of the impact of Marius on the Roman army.[109] Gruen's work on the last generation of the Republic argued strongly that the metanarrative that interpreted the last century BCE in terms of decline and recovery was more a matter of paradigm than evidence. In relation to the army, Gruen observed in 1994 that the 'idea that a gradual professionalization of the army since the Marian reforms does not easily meet the facts'.[110] Cadiou notes that not only Gruen but also Brunt has critiqued Gabba's view that a professionalized army led to a breakdown in Roman government.[111] Keaveney does not use the term at all to characterize the army during the last century of the Republic.[112] Eck, in his work on Augustus, does not refer to the army as professional, nor does his translator.[113] As we shall see later, men in Augustus' armies served for long periods due to renewals of his commands but with unforeseen consequences, transferring the problem of aged soldiers to Tiberius.

The application of the modern discourse of professionalism may be not only unjustified, it may be misleading too. Oakley's examination of Roman soldiers' engagement in single combat during the Republic leads him to find a tension between *disciplina* and the professionalism he assumes necessary for an army.[114] Lendon's examination of cultural context in which war was waged in antiquity provides him with a similar dilemma. Lendon argues that the armies of the Republic and Empire were quite different: 'a citizenry in arms' as opposed to 'a paid force of long-service professionals'.[115] The distinction ignores the fact that Roman soldiers had been receiving *stipendia* for almost 400 years before Augustus.[116] For Republican armies, apart from Caesar's command in Gaul, most commands were short-term, releasing soldiers when their commander's *imperium* expired or the command was transferred

The Myth of Professionalism in Rome's Armies before 14 CE 69

to a new magistrate as discussed earlier. Conditions of civil strife complicated this with the return of Sulla and his army from the East in 83 BCE as a prime example.[117] In Chapter 2, Cassius Dio's use of Maecenas to represent the change in Rome's armies during Augustus' time and the comments by Herodian provide further evidence that Roman armies were disbanded at the end of the command for which they were raised until Augustus' time. Whatever one may make of Roman armies from Augustus' time, there is nothing to support the assertion that earlier Roman soldiers were 'long-service' professionals, even if some repeatedly re-enlisted.

Returning to Lendon, wondering how the different Republican and Imperial armies could be similarly effective, Lendon finds, like Oakley whose study is of the Republican period, that *disciplina* provides the answer but only partly. Lendon's chapters on the Romans cover both the Republican and Imperial periods. He notes a characteristic tension between *virtus* and *disciplina*, but he writes that in Caesar's time, *disciplina* was dominant.[118] In spite of seeing that tendency, however, Lendon finds, like Oakley, that Roman *disciplina* did not function like modern military discipline, even for a Roman army that Lendon contends was 'professional'. In the siege of Jerusalem in 70 CE, for instance, he finds that the tension between *virtus* and *disciplina* remains.[119] Somewhat bizarrely, Lendon suggests that the tension may be resolved by seeing Roman soldiers, not as professional soldiers but as professional athletes.[120] His problem would be more easily resolved by abandoning altogether the use of the modern construct of professionalism to characterize Rome's soldiers, and accepting that *virtus* and *disciplina* shaped the behaviour of Roman soldiers in Vespasian's time, just as it shaped the Emperor Julian's behaviour later.[121] There really is no need to use 'professional' or 'professionalism' to characterize Roman armies to 31 BCE and even to 14 CE, as we shall see later. 'Professionals', 'professionalism', and the associated grammatical forms reflect historical developments and the growth of organizational theory in the last 150 years. The use of the terms has brought no clarity. If anything, it has demonstrated the risks of accepting too readily and uncritically the use of contemporary terminology for ancient realities. The extent to which it is helpful to describe the 'undying armies' in the Imperial period as professional is a task for another day, but Lendon's difficulty demonstrates that it too may be worthy of examination.

Part II

Rome's Armies Before and After Polybius

Chapter 5

Earliest Roman Armies

Introduction

So far, we have considered the history of 'the history' of Rome's armies to the death of Augustus. To a degree, we have considered not just how people in the past have looked at the Romans' military institutions, but also what we don't know about them. 'Professionalism' is an excellent example of the inappropriate application of modern views of the world to the Romans' armies in the later part of our period. We have seen that 'reform' is also an inappropriate concept to explain change because it is both unsupported by evidence and quite unlikely to have been implemented in the Romans' world. The alternative explanation of change is both simpler but frustratingly elusive: the idea that generals adapted to their circumstances and some of these adaptations became normal practice. In the chapters which follow, we will attempt to follow these changes, based on the ancient evidence, to define what we can know. For some, the reality which will be presented will be slender, even lean; we must accept that there is so much that we don't, even can't know.

Ab Urbe Condita (From the Foundation of the City)

What we may term the Roman polity always had soldiers of some kind from earliest times. All ancient political entities needed to defend themselves from enemies or have a capacity for taking land, people or goods from neighbouring states. There are descriptions of Rome's earliest armies in the writings of Livy and of Dionysius of Halicarnassus. The representations of these armies were consistent with each author's literary purposes in writing their works. Any consideration of what we can know about Rome's armies before the First Punic War should begin with a consideration of Livy's comment in Book VI.1 where he notes the difficulties of knowing the truth about events up to 700 years before his time.[1] Livy wrote that he had to rely on commentaries of the pontiffs (priests), and various other documents; no less than fourteen different sources were listed.[2] Surprisingly with the dearth of evidence which Livy considered reliable, there is a notable consistency in the historical sources of early Roman history, including between Livy and his Greek contemporary, Dionysius of Halicarnassus.[3] This can be explained by the existence of an accepted narrative, even a metanarrative, when Livy and Dionysius wrote their histories.[4] This is important to note. The Romans had a view of their past which included their military institutions. As far as we know, Roman historical writing began with Fabius Pictor in the third century BCE.[5] Pictor wrote in Greek for a Greek speaking audience.[6] There were Roman aristocratic family histories, though they were notoriously unreliable according to Cicero.[7] In the first century BCE, Q. Claudius Quadrigarius wrote a history of Rome beginning in 390 BCE because he believed that all earlier records had been lost.[8] The question which must be faced is to what degree can the accepted ancient narrative be trusted today. The sources are poor.[9] Some scholars have dismissed the views of the past in Augustus' time as mythic inventions which were aimed at ideological justification or, at least, to address contemporary perspectives.[10] Other scholars believe that even with the difficulties, it is possible to discern the development of the Roman state from its earliest times.[11] Even knowledge of military equipment is very limited due to a lack of sufficient literary or archaeological evidence.[12] However as a starting point, there appears to have been a period of what might be called state formation in the eighth and seventh centuries BCE.[13] By around 509 BCE, the Republic was being dated by the Fasti, calendars which recorded important dates which were commemorated.[14]

The nature of the earliest Roman armies is as much a matter for conjecture as the nature of the political institutions.[15] In the discussion of the reform paradigm, 'Servian reforms' were mentioned. The root of the claims for their existence is to be found in the statements in ancient texts which do claim that a king of Rome, Servius Tullius, was responsible for some changes related to the Romans' armed forces. The texts which describe the earliest armies are Livy's early books and Dionysius of Halicarnassus' history, both dating from the late first century BCE.[16] There are references in other authors and there is a certain amount of archaeological evidence.[17] Livy presents Servius Tullius as performing a role in the organization of people, equivalent to that played by another king of Rome, Numa Pompilius, in the organization of Roman religion.[18]

The descriptions of an organization created by one man need to be read with caution. Both Livy and Dionysius describe an army which combined people with different weapons into one force. The key text from Livy, which describes the changes made by Servius Tullius, states that the citizens were divided into five classes, each of which had different weapons. The first class sounds much like Greek hoplites but the second class has the *scutum*, the Roman rectangular shield, rather than the *clipeus*, a round hoplite-style shield; the *pilum*, the Roman heavy javelin, is not mentioned. The lowest census class was excused from military duties. Livy also presents Servius Tullius as undertaking the organization of the cavalry. And finally, Servius Tullius organized the people into the tribes which were to survive for centuries as a basis for political organization.[19] Dionysius of Halicarnassus' description of the changes instituted by Servius Tullius is in substantial agreement with Livy's account. There is a similar combination of organization for war and for peace.[20]

These texts raise three key questions. First, did these changes happen? Secondly, did these changes occur all at once? Thirdly, was a sixth century BCE king named Servius Tullius responsible? The historicity of the existence of the institutions, as opposed to the military organization, said to be created by Servius Tullius is not open to serious question as they existed later.[21] But similarly, the census arrangements ascribed to Servius Tullius cannot be his, probably being no older than the fourth century BCE, and reflected political, not military arrangements.[22] The military origins of the *Comitia Centuriata* should not be doubted either, given its title and its later powers. Quite apart from the military terms used in its organization, the *Comitia Centuriata* described by Livy and

Dionysius bore the marks of a military background.[23] For example, to summon the assembly, the term was '*convocare exercitum*' ('assemble an army'). The assembly was described as '*exercitus urbanus*', ('town army').[24] Dio records that meetings were held under circumstances which would allow the army to fight with as little time lost as possible.[25] However, the assumption that there was a single legion of 6,000 men more likely reflects that the original meaning of *legio*, i.e. legion, was a gathering or a levy.[26] At some time, this assembly almost certainly had a military function, though it does not seem to have had a solely military function.[27] It is impossible for us to know whether the description of the changes includes only, or even mostly, those things which were done at one time, but it is doubtful. Given Livy's comments about the paucity of evidence in the available sources and both his and Dionysius' remoteness from the events, it is difficult to have much faith in the detail they provide. Livy characterizes the changes as for the benefit of the state: '*res saluberrima*' ('a very healthy thing'), not as reforms.[28] It is true that the ancient view was that one man was responsible for the changes, i.e. the lawgiver, but lawgivers were a common device for explaining an ancient state's institutions when their origins were obscured in the mists of time.

Dionysius' account with its reference to different authors shows that he at least saw a need to clarify what occurred by comparing different accounts: Fabius Pictor, Cato, Vennonius and Piso.[29] This may not have been easy. Livy in Augustus' time frankly acknowledged the difficulties of finding information about this early part of Roman history, as we have seen.[30] Dionysius' description lays stress on tribes and the inclusion of outsiders as citizens. This approach recalls Herodotus' treatment of the changes credited to the Athenian Cleisthenes, another lawgiver.[31] It also is important to consider why the central feature of the changes credited to Servius Tullius took place, i.e. why was the census instituted? The links with the Greek historical tradition are not confined to Cleisthenes, there are Solonian echoes of census classification, also from Athenian history.[32]

Setting aside for a moment the issue of one man's responsibility, how likely is it that Roman armies could have been as Livy and Dionysius described them? The change, which modern authors have detected, focuses on the adoption of hoplite tactics, organization and equipment in the first census class by the Romans in the sixth century with different equipment in the census classes.[33] From as early as we have information, ownership of land appears to have been a mandatory requirement for being enlisted in a Roman army. The *assidui*, that is 'settled landowners', were qualified. The *coloni*, tenant farmers, were given land in colonies, taken from other peoples.[34] There can be little doubt that what we see as hoplite-style armour was used in ancient Italy.[35] Dionysius specifically describes the equipment of the first census class as including Argive shields, a sure indication that he believed them to be hoplites.[36] A number of modern writers claim that the hoplite style was the sole form of infantry combat prior to the Battle of the Allia, but assumptions about the homogeneity of hoplite armies in this period are open to question.[37] Archaic Greek warfare down until possibly 650 BCE was less a matter of homogeneous hoplite phalanxes than mixed formations such as are attributed to the Romans in Servius Tullius' time.[38] Even later Greek warfare was not necessarily characterized by close formation fighting only.[39] It is quite likely that older military practices survived on the Italian fringes of the Greek world.[40] The Certosa Situla in the Museo Civico in Bologna (see Figure 7), dating from circa 600 BCE, only a little earlier than the probable date of Servius Tullius' reign, shows what appears to be an army composed of troops equipped in different ways, just as Dionysius and Livy describe Rome's army of this period.[41] If we accept that Greek warfare, prior to this period in Italy, involved mixed formations, such as seem to be depicted on the Certosa Situla, then arrangements described by Livy and Dionysius are not *a priori* improbable and may be accepted as having a factual basis, but it is impossible to say whether the historical Servius Tullius was responsible, let alone to divine the circumstances that might have led to any arrangements he made. It is possible that Servius Tullius adopted military arrangements from neighbouring Italian peoples. It has been argued that the depiction of Italiote peoples, particularly Etruscans, in hoplite-style panoplies was an artistic claim of distinctiveness from the Romans.[42] Etruscan grave goods suggest that wealthy Etruscans were equipped similarly to wealthy

Romans.⁴³ The claims of differences in military equipment between the Romans and their Italian neighbours may well be exaggerated.⁴⁴ The stumbling block is really Livy's admission that he found it very difficult to know what had happened in a past that was distant from him. Given the tradition of law-givers that was so common in accounts of the origins of classical states and peoples, it is difficult to have a great deal of confidence in the Servian/Tullian army changes, let alone to see them as reforms, although some scholars remain attracted to the idea of a single author.⁴⁵ It is at least as likely that military developments occurred through the actions of one or more individuals, in process of development.⁴⁶ Scholarly opinion inclines to the view that in earliest Roman history, there was simply a single *classis*, classification, of those who had the equipment to fight as opposed to those who lacked it. Even the idea that the Romans fought as hoplites is open to question, undermining ancient accounts; the Romans do not appear to have fought as hoplites in the fourth century BCE.⁴⁷ If we abandon the idea of a single author of Rome's earliest military arrangements, what is the alternative?

A feature of early Roman warfare is that it appears to have been clan based. The evidence is found partly in historical and partly in archaeological texts. In 1977, an inscription to *Mater Matuta*, an Italian goddess, was found at Le Ferriere, ancient Satricum. The text refers to a dedication by the *suodales* (companions) of Poplios Valesios dated to circa 500 BCE. Poplios Valesios has been identified as Publius Valerius Publicola, Roman consul in 509, 508, 507 and 504 BCE. The *suodales* are thought to represent a war band.⁴⁸ A similar example may be found with the migration of Attus Clausus and his *clientes*, followers, to Rome in 504 BCE.⁴⁹ A further example may be in the attempt by Appius Herdonius and his armed group to take control of Rome.⁵⁰ Similar activity appears in Etruria where 125 helmets were found in Vetulonia, which may be relics of a clan army. The Sperandio sarcophagus from Perusia (modern Perugia), dating from the same period, commemorates what appears to be a raid for cattle and prisoners.⁵¹ If these interpretations of the evidence are correct, it is hardly surprising that neither Livy nor Dionysius emphasize such activity in their narratives 500 years later when the Roman state was very different. Clan warfare of raiding and cattle stealing is much more like the seventeenth and eighteenth century Scottish Highlands than our image of the Romans but it may provide a more accurate representation of the Romans' earliest armies. The organization of the legions, as we recognize them, seems to have appeared in the latter half of the fourth century BCE.[52] Yet the interest in what we would see as ill-disciplined military behaviour like clan raiding continued in an enthusiasm for individual combat which required regulation.[53] This is yet a further reminder that we should not apply modern views of appropriate military behaviour to the Romans.

We can accept that the city of Rome was sacked by a Gallic force following a defeat at the Allia River around 390 BCE.[54] The event had a substantial impact on the Romans. One Roman response was the construction of extensive fortifications, often referred to as the Servian walls.[55] The stone found in surviving sections of the walls shows that they were not built before the early fourth century BCE, well after Servius Tullius.[56] Using some stone from the territory of the destroyed Veii, the fortifications enclosed the large area of 427 hectares. This surpassed the enclosed area of any comparable Italian city for some centuries.[57] It is likely that the walls were expanded over time, as required. Walls such as these were to serve the Romans well, particularly in Hannibal's war, providing a secure political base from which armies could be raised. These walls served a military defensive purpose, but we have little idea of how they were manned and managed. They have been linked to the Servian changes which may have used the four districts of the city both for levying troops and defence.[58] It is possible that Gallic incursions into central Italy had a broad impact on Italian warfare. The adoption of oval shields, similar to Gallic shields, replacing round shields in Italy and the use of Gallic-style helmets of Etruscan manufacture suggest adaptation to the circumstances the Romans and other Italian peoples faced.[59]

Maniples

Livy records that the Romans began to provide pay (*stipendium*) to their soldiers just prior to the siege of Veii c.396 BCE.[60] *Stipendium* may have Etruscan equivalents.[61] It has been assumed that the tax known as *tributum* was initiated to fund the soldiers' pay but this may not be so.[62] The Romans do not appear to have had a treasury to coordinate the collection of *tributum* or the payment of *stipendia*.[63] However, if the need to maintain troops besieging Veii is the reason for the initial collection of *tributum* to pay soldiers, this is another example of contingent adaptation to particular circumstances; in this case, the adaptation became regular practice.

At some time in the period between the Battle of the Allia River and the war against Pyrrhus, 282–275 BCE, the Romans began to use an organization based on maniples and equipment suitable for a more individual style of fighting.[64] These changes to the Roman army are described as the 'Camillan reforms', named after M. Furius Camillus (c.448–365 BCE), although not by all scholars.[65] Coussin accepts that changes to Roman equipment should be attributed to Camillus, though he does note some voices dissenting from the acceptance of a single reformer.[66] Parker notes the improbability of the changes attributed to Camillus being implemented 'all at once'.[67] Miller has questioned whether the reforms existed at all.[68] Keppie, who uses reform as a synonym for change, uses the term to describe changes to the army in this period, although he does not accept that Camillus was entirely responsible.[69] It would seem that there are few scholars today who still believe that the so-called Camillan reforms occurred, though many include changes that created the new legions as reforms. Such views actually undermine the assumptions implicit in reform, discussed in Chapter 3. Cornell however maintains that it is possible that at the end of the fifth century BCE, a combination of a political reorganization, the introduction of pay for soldiers, and different tactics and equipment including a change of shields may have occurred.[70]

The starting point for any consideration of the changes associated with Camillus must be Dionysius of Halicarnassus' account of Camillus' activities. His text refers to soldiers' equipment and organization in a speech used to represent Camillus' encouragement to his soldiers.[71] The Roman advantage, according to Dionysius' representation of Camillus, is in superior equipment. Later, Camillus contrasts a smaller, trained army with a larger untrained one without any hint that the Roman force was the result of a recent change.[72] A slightly different account involving Camillus is found in Plutarch's text, where Plutarch explains the measures which Camillus undertook to help his men win against the Gauls.[73] These things include helmets and metal-rimmed shields and the use of a heavy javelin as a spear. Plutarch's text reads as an account of the directions of the elderly general to his soldiers to fight a particular enemy, though some of the advice is difficult to understand. Camillus instructs his men to catch the Gauls' long swords on their spear shafts to blunt them. It is unclear how this would be effective.[74] Polybius describes identical tactics as being used against the Gauls by Flaminius in 223 BCE, suggesting that it may be a literary trope.[75] Even if we give Plutarch and his source the benefit of the doubt, we still have Camillus directing his troops against a particular enemy.[76] There is nothing to indicate recent changes, institutional changes or even to suggest that these changes were meant for armies other than Camillus'. It remains that there were probably changes to Roman equipment and organization in the period between the sixth Century and the First Punic War. Livy (8.8) describes a legionary organization which he dates to the landing of Alexander of Epirus in Italy in 341 BCE. However, Livy makes no mention of Camillus in relation to the organization of the troops. A new organization can be explained by Camillan reforms, but it is speculative and unlikely and almost no one accepts them as a package attributed to one man. If Dionysius' account of Camillus' speech has any historical value, it reflects a contingent adaptation to the circumstances of a war against a particular enemy as a particular time. This is supported by the tradition in the *Ineditum Vaticanum* of the Romans' adoption of different weapons and organization from their enemies.[77] Even here, caution is needed. Representations of Italian military equipment in the fourth century BCE show many of the elements of hoplite warfare: circular shields (*clipei*), greaves and body armour. Even Samnites are shown in this fashion. Multiple spears and an interest in individual combat suggest that it is unwise to assume that we understand Italian warfare in this period.[78] In sum, there is no substantial agreement on the dating of changes in Rome's armies in the fourth century.[79] Nevertheless, the Romans believed that changes did occur in this period, associated with Camillus, the Gauls and the Samnites.[80]

Roman armies using maniples were probably the result of years of adaptation to the military realities of warfare in Italy, although it is possible that the Romans learnt of the Samnites maniple style of warfare when they were allied with each other at the Battle of Veseris in 340 BCE.[81] There is even some evidence that the Etruscans used a manipular organization in 283 BCE.[82] The later army described by Polybius arrayed troops in three lines of *hastati*, *principes* and *triarii*, each of which was composed of *manipuli*, ('maniples' or 'handfuls'), although we have no evidence when these arrangements began.[83] Livy's record of the soldier's oath includes the commitment not to leave the battle line unless to, among other things, strike an enemy.[84] This implies a much looser formation than Greek or Macedonian battle lines where such an action would make no sense.[85]

Polybius tells us that in his time, all but the *triarii* used a javelin, the *pilum*. The use of the term *hastati*, spearmen, for one of the two other types of close-fighting infantry suggests that by Polybius' time the term reflected an older usage.[86] Livy in Augustus' time accepted that their equipment had changed over time, based on their experience in war, i.e. they adapted to their enemies.[87] The Samnites were cited as the source of the Romans' rectangular shield, the *scutum*.[88] There may be some anti-Samnite bias in Livy, who rejects the Samnites as the source of the long Roman shields; both Dionysius of Halicarnassus and Sallust disagree with Livy.[89] Literary evidence strongly suggests that the Romans believed that the *scutum* and the *pilum* came from the Samnites or the Gauls.[90]

Polybius describes Roman armour as helmets, greaves and either mail for the *triarii* or small breastplates which covered the chest. Except for the mail, equipment similar to this has been found in Italian tombs.[91] The earliest chainmail has been found at Ciumesti, Romania, dating from third century BCE. Depictions of mail dating from second century BCE have been found in the lower Rhone valley suggesting that the Romans would have had access to it.[92]

The *scutum* is probably linked to the development of maniples because it suited an individual style of combat.[93] Maniples don't assume a continuous hoplite style battle line but more of a series of clusters. Unlike hoplites who used each other's shields as cover, a soldier with a *scutum* fought singly, not depending on his fellows as Polybius tells us.[94] The shape of the shield may have been adopted from Gallic models.[95] As for the *pilum*, the Roman heavy javelin, weapons with a small head and a long shaft have existed from early times and it would appear that there are significant similarities between Etruscan, Roman and Gallic javelins, and the Etruscans seem to have used their version with the hoplite panoply.[96] An obscure reference in Festus to a phrase in the hymn of the ancient Salian priests suggests that Roman use of the *pilum* was very old.[97] Even the Romans' habit of constructing temporary marching camps may have arisen from Samnite practice, although there is disagreement among ancient writers.[98]

In terms of training, these Roman armies were probably a militia whose members had little weapons drill and were expected to learn over a series of campaigns in each of the temporary armies in which they served.[99] While there is much that we would like to know about Roman armies in this period, there is sufficient evidence to suggest that the Romans used equipment, at least some of which, would have been copied from enemies and friends. There is simply no need to postulate reform to explain this; contingent adaptation leading to accepted practice is both simpler and more consistent with the slender evidence.

The nature of Roman military leadership is an interesting aspect of Rome's armies but little is known of its evolution. This early state was almost certainly oligarchic. Clan-based raids existed with full levies of Roman and Latin allies with armies led by aristocrats with the title of *praetor*.[100] The later highest magistracy, the consulship, is likely to have had a military origin, dating from around 367 BCE. From 405 to 367 BCE, the Romans appear to have elected six military tribunes with the powers of consuls.[101] The Latin term for legal authority to command troops in Cicero's time was *'imperium'* while the power of a non-military magistrate in Rome was called *'potestas'*, a term which may be older than *'imperium'*.[102] The distinction between the two kinds of authority is significant because it limited military commanders' quite arbitrary powers over soldiers.[103] Quite unlike their modern counterparts, Roman generals, from as early as we have evidence, needed to consult the gods before committing their armies to battle. The responsibility of taking the auspices (*auspicium*) was taken very seriously because war put the Roman state potentially at risk.[104] Valerius Maximus, writing in Augustus' time, has the moralizing tale of Publius Claudius Pulcher, the Roman commander at the naval Battle of Drepana in 249 BCE, who, frustrated that the chickens used to take the auspices would not show the gods' support for battle by eating, flung the birds into the water to drink instead; he lost the battle.[105]

Apart from the obligation to take the auspices, generals' *imperium* gave them considerable freedom and independence provided they complied with a number of requirements. The first was that a general could not exercise *imperium* inside the *pomerium*, the sacred limits of the city of Rome. The second was that *imperium* could only be exercised within a general's defined sphere of activity, his *provincia* which later came to refer to a geographical entity, i.e. a province. The third was that a general could be subject to prosecution once he laid down his *imperium*. The fourth limitation allowed a popular assembly to revoke *imperium*.[106] Nevertheless, a Roman commander could make a local decision which, in the case of Ap. Claudius at Messana in 264 BCE, could commit the Romans to a war with Carthage which lasted twenty-three years.[107] Once battle was joined, a commander had to rely on the training of the soldiers, the quality of the *tribuni militum* (tribunes of soldiers), and the centurions.[108]

Any discussion of Rome's armies in the period prior to the First Punic War is incomplete if it does not consider the allied forces, assuming that Polybius' account of the composition of Roman armies in his time applied to the earlier period.[109] The Romans' capacity for incorporating troops from different polities, speaking different languages and possibly using different tactics, is notable.[110] The readiness of the Romans to enfranchise their defeated enemies is surprising to modern eyes which do not appreciate that these former enemies, even *'cives sine suffragio'* ('citizens without voting rights'), provided both manpower and taxes (*tributum*).[111] However, this practice was not used much beyond Latium.[112] Because *tributum* funded *stipendia*, there was a disincentive to use too much manpower. Allied troops, the *socii*, provided by their own communities, provided no such disincentives.[113] We know almost nothing of the composition or equipment of these allied forces. As we shall see below, Polybius's description in Book VI implies that Rome's Italian allies were equipped in a fashion similar to that of the Romans without explicitly saying so.[114] However, few consider what the Romans' alliance offered the allied soldiers who served with the Romans. The answer must surely lie in booty, a purely economic motivation, and the Romans' aristocratic military ethos which valued achievement in war so highly.[115] There is no reason to believe that the aristocracies of Rome's allies had a different view.[116] The Romans required their allies to supply troops according to the *formula togatorum*, the register of those who wore the toga of adult males.[117] Polybius states that food was provided to both allied and Roman troops but the Romans had the cost deducted from their pay.[118] Polybius' description of the Romans' punctilious attention to the distribution of booty suggests that it was distributed to allies and Romans.[119] The successes of Roman armies provided wealth to both Roman and allied soldiers and their communities.[120] It is reasonable to assume that Roman troops and their allies were motivated by booty from the foundation of Rome.

Chapter 6

Polybius and his Roman Armies

On Firmer Ground: Polybius

The most comprehensive surviving account of the Romans' military institutions before the death of Augustus comes from Polybius and it is through him that we can claim to know how the Romans' forces were raised, equipped, and organized in his time and much later. Polybius account is the earliest reliable literary source on Rome's armies.[1] Livy's work contains a description of Rome's early armies as we saw in the last chapter.[2] Dionysius of Halicarnassus, Livy's contemporary, provides a similar account.[3] At best, Livy's description is a mine of Latin vocabulary related to Rome's armies; at worst it is impossible to untangle.[4] Polybius did not write about Rome's armies in his time because of a particular interest in military institutions, although he is one of the few ancient authors to engage with such matters.[5] Unlike Caesar's later commentaries, what Polybius provides is not inferential or an unintended outcome of other purposes in writing. In fact, Polybius' account of the Roman forces is found in his description of the Romans' *politeia* (πολιτεία), a word often translated as 'constitution' but with a broader meaning encompassing all of a nation's institutions. This account in Book VI is consistent with Polybius' purpose in writing, previously stated in Book I:

> For who is so worthless or indolent as not to wish to know by what means and under what system of polity the Romans in less than fifty-three years have succeeded in subjecting nearly the whole inhabited world to their sole government – a thing unique in history?[6]

Polybius, a well-educated Greek hostage in Rome in the second century BCE, wrote to explain to a mainly Greek audience how this Italian power from the periphery of the Greeks' world was able to defeat or humiliate all of the Macedonian successor states and even defeat armies of the Macedonian king.[7] Polybius wishes his countrymen to see the world as it truly is including the possibility that the Roman polity may have been better in the past but he is not without cultural prejudice.[8] Polybius states that he knows Romans might look at his work but he maintains that he will be respected for telling the truth and he appears to have had a sound knowledge of Latin.[9] Polybius has been criticised for an overly systematised depiction of the Romans' armies.[10] Such criticisms from a perspective of 2,200 years may be treated with scepticism as they second guess Polybius' literary judgement and are based on very limited knowledge today. For our purposes, we will concentrate on what Polybius, a militarily knowledgeable observer, says about the Romans' military arrangements and even what Roman soldiers looked like.[11]

Polybius was not Roman, although, as a hostage and later perhaps a guest, he seems to have spent most of his adult life with Romans, at least some of whom were eminent men like Scipio Aemilianus. While the date of his birth is uncertain beyond the end of the third century BCE, he came from Megalopolis in what is today Greece. Megalopolis was a principal city in the Achaean League in southern Greece. Polybius' father was Lycortas, a prominent politician. Polybius' Achaean background emerges when he comments, to his Greek audience, on the merits of the Achaean League and of its leading man, Aratus.[12] Polybius was destined to become an important political figure in the Achaean League until he was given as a hostage to the Romans in 170 BCE; by that time, he had already been designated as an ambassador to Ptolemaic Egypt, although the death of Ptolemy V prevented the trip being undertaken. Polybius was a hipparch, a commander of cavalry, in 170 BCE, before being sent to Rome as a guarantee of good Achaean behaviour.[13] The date was significant, marking a decline in relations between the Romans and Perseus, the king of Macedonia.

The world of Polybius, the Greek from Megalopolis, was dominated by the shadow of Alexander the Great. The defeat of the combined forces of Boeotia, Athens, Chalcis, Epidaurus, Megara and Troezen at Chaeronea in 338 BCE was a major turning point in Greek history. From being a land of independent city-states, often in competing alliances since earliest recorded history, it was from that time, until the Romans, dominated by a previously peripheral state which controlled the Greek states through Macedonian garrisons in Demetrius, Corinth and Chalcis, 'the fetters of Greece'. The once-dominant Spartans had already been defeated by the Thebans at Leuctra in 371 BCE and Mantinea in 362 BCE. Following his victory at Chaeronea, the Macedonian king, Philip forced the Greek states, except Sparta, into a League of Corinth. The previously independent states were compelled to live peaceably and to support the Macedonians in the campaign against the Persians. Philip's successor, Alexander, conquered the Persian Empire but his death in 323 BCE saw the Eastern Mediterranean split into what are referred to today as the Macedonian Successor states. In Polybius' time, after a century of wars, these were Macedonia, Ptolemaic Egypt and the Seleucid Empire including Mesopotamia and Syria. There were a number of smaller states like Pergamon and confederacies like the Achaean League. This world was very different to the Classical Greek world of Thucydides or Xenophon. It acknowledged the power of the great Macedonian Successor kingdoms based on the military system which we associate with Alexander the Great.

For Polybius, a man educated to be a leading figure in the Achaean League and trained in contemporary Macedonian-style Greek warfare, an explanation of the Romans' success using a very different style of warfare is the major theme of his historical work. Polybius speaks quite directly to his Greek audience, for example, when he seeks to correct Greek geographical misconceptions or when commenting on earlier practices of Greek inter-state relations.[14] When we read Polybius' account of Rome's political, military and social arrangements in Book VI, we ought to imagine a conversation with a contemporary Greek speaker, quite familiar with the Macedonian style of warfare and Greek political structures. We will look in vain for details which Polybius did not provide.[15] And we should not be surprised if his treatment of his subject seems overly simplified, although it is worthy of reflection as to the degree to which our modern experience of highly structured modern armies exacerbates the problem.[16] Polybius simply did not see some details we seek as relevant to what he felt his contemporary Greek speakers needed to hear, keeping in mind that most would have heard his work read out loud, probably by an educated slave. It is fortunate that we possess a number of surviving ancient works which describe the Macedonians' military arrangements. These date from the early Roman Empire following the death of Augustus. Modern writers see this as a literary fashion, known as the Second Sophistic, which saw what appears to have been a renewed interest in Greek philosophy. However, other interests demonstrated a confidence in Greek cultural values including military organization. We should not forget that during the Roman imperial period, Greek states often maintained military forces which were deployed with Rome's forces including against the Jewish rebels in 70 CE.[17] One of the authors who celebrated Macedonian military arrangements was Arrian, to whom we are indebted for his detailed account of Alexander's campaigns. However, Arrian was also a Roman official and commander of Roman forces. As previously mentioned, he left us another work in Greek, *Tactics against the Alans*, describing tactics which are more recognizably Macedonian than Roman. We also have the works of Aelian and Asclepiodotus to whom Seneca, writing in Nero's time, may have referred.[18] Together, these give us some understanding of how Macedonian armies operated. As a result, we can engage with Polybius' contrast between the Romans' military arrangements and those of his Greek-speaking audience. However, we cannot expect Polybius to provide what we would take to be a comprehensive description of Rome's armies for our time; that was not his aim.

Polybius' Roman Armies

Reading Polybius' description of Roman military arrangements as information for people familiar with Hellenistic military practice provides an otherwise missing context for his treatment of the Romans. Polybius is at pains to emphasize that Roman citizens were required to make themselves available to be enlisted annually.[19] The difference here to Hellenistic practice is that annual campaigning was not common. Hellenistic rulers maintained royal armies but raised additional forces only when needed.[20] The military implication here is that Roman soldiers could be quite experienced due to years of service, like Livy's Ligustinus whom we met earlier.[21] Polybius reinforces this point by detailing the care which the Romans took to ensure that each legion had the same proportion of men both experienced and inexperienced men. Polybius is clear that Roman allied troops from Italy were selected in a similar careful manner, emphasizing the consistency in the quality of Roman troops.[22] We may suspect that Polybius is suggesting that other nations did not achieve this degree of consistency in the quality of their manpower.[23]

Polybius explains how the Romans annually form each legion from four different classes of soldier: 1,200 *velites*, 1,200 *hastati*, 1,200 *principes* and 600 *triarii*. Each of the last three classes are divided into ten maniples, each commanded by two centurions and two *optiones*. Between the thirty maniples, the 1,200 *velites* are distributed, forty per maniple, where there were officers to command them. Because there were no officers detailed for the *velites*, they must have been integrated into the maniples and not organized separately.[24] This is consistent with the remains of a second-century-BCE Roman camp in Spain.[25] He further states that *velites* had a quite effective shield, a sword and javelins. The implication from this equipment is that *velites* are not simply skirmishers: these men were expected to fight hand to hand.[26] This distribution of the *velites* brings the strength of the maniples of *hastati* and *principes* to 160 men each and 100 men for the maniples of the *triarii*; 300 cavalry are added from a list managed by the censors after the selection of the infantry.[27] Some 300 cavalry were assigned to each legion, although Polyius also refers to 200 as the normal assignment.[28] While such legions numbered 4,400, legions could have 5,000 if needed.[29] These legions are clearly age based. The *velites* were the youngest and financially poorest and then, ascending in age and wealth, the *hastati*, *principes* and finally the *triarii*.³⁰ Apart from the *triarii* and the cavalry, 3,600 men in each legion are equipped with missiles, suggesting that the Roman infantry did not rely in hand-to-hand combat alone.

The description of how the Romans were armed and equipped shows Polybius' intent that his readers should know what the Romans called their soldiers where that differed from Hellenistic practice. The Romans' weapons are described in detail. While Roman *velites*, that is the light-armed troops, are referred to as γροσφομάχοι (*grosphomachoi*, i.e. 'javelinmen'), a term which would be easily understood by Greek readers, he uses Latin terms for the more heavily equipped Roman infantry: *principes, hastati* and *triarii*. Polybius uses Greek terms for their weapons: the spears of the *triarii* are δόρατα, (*dorata*), and the Romans' heavy javelins, the *pila*, are called ὑσσοί (*hyssoi*), a heavy hunting javelin or spear; clearly Polybius did not see *pila* as particularly noteworthy. Numerous examples of *pila*, corresponding to Polybius' description, have been found in archaeological sites from Spain to Slovenia.³¹ When describing the *velites*, he details the size of their *parma* shield and their javelins.³² Asclepiodotus' description of Hellenistic light-armed troops describes them as missile troops with a light shield called a πέλτη (*pelte*), different to the shields of the *velites*. Polybius tells his readers that *velites* had helmets, unlike Hellenistic light infantry, and he also describes the nature of the javelins carried by *velites*.³³

When Polybius moves on to describe Roman heavy infantry, he notes that they wear a panoply, i.e. armour, but he distinguishes the Roman panoply as being different from what his readers would assume, based on Hellenistic practice. The detail is extensive although some modern authors have questioned it.³⁴ The Hellenistic audience would have noted that, unlike the Hellenistic heavy troops who fought in a phalanx, as described later by Asclepiodotus, Roman heavy infantry used missiles rather than pikes or spears and relied on a 'Spanish' sword which could both cut and thrust and a large shield.³⁵ Archaeological evidence supports Polybius' description.³⁶ The helmet in Figure 10 is similar to that described by Polybius. The Pydna Relief (Figure 11) shows what appears to be a figure with a shield much like that described by Polybius.

Polybius ascribes the Romans' advantage in combat to the fact that Roman soldiers could turn individually to face an enemy in any direction, confidently protected by a large shield and a robust sword.[37] For modern readers, Polybius' description of Roman equipment may be viewed as a reason for the Romans' success but we should be cautious. Polybius does not claim that Roman equipment was the reason for their success, although it is a factor. The modern temptation to see a technological determinism ought, in Polybius' view, be avoided.[38] His claim is that Rome's success was due to its *politeia*, which gave them what we might loosely call a moral advantage.

Following his description of the very different Roman equipment, compared to Hellenistic practice, Polybius then describes how the Roman troops are commanded and led. He used Greek terms for some officers. He refers to centurions as ταξίαρχοι (*taxiarchoi*), corresponding to Asclepiodotus, who describes *taxiarchoi* as commanders of 128 men but notes that they were also commonly known as commanders of 100.[39] Polybius refers to men appointed by the centurions as οὐραγοί (*ouragoi*), ie 'rear men' or 'file closers' in a phalanx. In later Roman history, *optiones*, literally 'chosen ones', are found in Rome's armies.[40] For the different subunits, Polybius uses the Greek terms τάγμα (*tagma*), σπεῖραν (*speiran*) and σημαίαν (*semaian*), i.e. non-specific groupings, bodies, followers of a flag or standard. He notes that the leaders of these groupings appoint two men as standard bearers, (σημαιαφόροι, *semaiaphoroi*), equivalent to *vexillarii* in Latin. Polybius explains this by the need to always have commanders present to lead the troops. He follows this with the observation that the Romans appointed men as centurions who will hold their ground rather than men who will show initiative and take risks.[41] There may be an implication here that his Greek-speaking audience expect leaders to be more daring, recalling Thucydides' descriptions of men like Brasidas. Polybius' description of the Roman cavalry details differences from Hellenistic practice. There are three brief treatments of cavalry in Arrian, Aelian and Asclepiodotus; they are very similar, suggesting that either they all drew on a common source or the first two used Ascelpiodotus as their source. All describe cavalry formed in rhomboids or wedges. Polybius and his Greek audience would have been very well aware of such practices. Polybius does not comment on Roman cavalry formations but, instead, takes the opportunity to explain to his readers that the Romans adopted Greek equipment when they found that to be superior, a contingent adaptation to circumstances.[42]

Before explaining how the Romans march and encamp, Polybius describes how the Romans manage their allies in Italy using twelve πραίφεκτοι, i.e. *praefecti*, prefects. He transliterates the Latin word because there was no equivalent for his Greek audience. He does the same thing in the following sentence in explaining that the Romans select the best of the allies' cavalry and infantry: ἐκτραορδινάριοι, i.e., *extraordinarii*, meaning 'picked' or 'selected', although he does not use the term later.[43] Polybius tells his audience that Roman armies have the same number of Roman and allied infantry but that the allies supply three times the number of cavalry. The infantry are divided between a right and a left wing.[44] We may only now surmise why Polybius included this detail. It may have been for completeness but it may also be because the Romans managed their allies differently to Hellenistic practice. Consequently, Polybius felt that his Greek audience needed to understand why these allies mostly remained loyal to the Romans during the Hannibalic war. Polybius does not indicate that the equipment of the allies was different from that of the Romans.[45]

The section of Polybius' text which follows describes how the Romans encamped, even noting that in Flamininus' campaign in northern Greece, which culminated in the Roman victory over the Macedonians at Kynoskephalai in 197 BCE, Roman soldiers could carry stakes for a palisade because they were not encumbered with a pike and were able to carry their shields with a strap over their shoulders.[46] The detail on Roman encampment practices is extensive, permitting someone with an interest in such matters to copy or modify the Romans' arrangements. This appears to be Polybius' motive for including the section, as he says at the end of Chapter 27 of Book 6. He follows up the details with a brief discussion of the benefits of such a camp layout, more detail on how it is scalable for larger forces, the soldiers' responsibilities for their tribunes, and how it is fortified and secured.[47] Polybius then speaks directly to his Greek audience on the superiority of Roman methods compared to Greek encampment practices.[48] In Book 18, Polybius, in an aside, explains the advantages of the Roman's choice of stakes to fortify their camps.[49] Modern scholars have debated what kind of camp Polybius was describing, not always profitably.[50] Archaeological evidence from Spain shows that in the second century BCE the Romans built camps of different sizes. The stone-built camps around Renieblas and Numantia have been extensively excavated and they are consistent with Polybius' account of how the Romans encamped, given that he describes a camp for four legions, i.e. a double consular army, and that many of the Spanish camps served smaller armies.[51]

Apart from this extended description and commentary on Roman military practices, Polybius makes a number of other observations about the Romans' armies elsewhere in his text. He claims that the Romans gained their military expertise from their struggles against the Celts and Samnites.[52] This is consistent with what we have earlier surmised about the origins of Roman equipment and even tactics. Polybius shows an interest in some issues which may seem peripheral to us but were probably not to his Greek readers. He notes the problems with the employment of mercenaries, an issue of interest to Hellenistic polities which employed Celtic and other mercenaries.[53] Polybius is clearly impressed with Hannibal as a general, not just for his victories but for how he managed his polyglot army.[54] Polybius shows his interest in military affairs ranges far further than the Romans. He devotes part of Book 9 to what is required of a commander.[55] He tells his readers that the Roman and Carthaginian cavalry fought dismounted, 'barbarically' (βαρβαρικήν), without the manoeuvres which his Greek audience might have expected.[56] Based on the result of the Battle of Cannae, Polybius, a former cavalry commander, notes that it is preferable to have an overwhelming superiority in cavalry but take fewer infantry.[57] Later, Polybius provides advice on cavalry manoeuvres perhaps reflecting Hellenistic military practices based on Philip and Alexander of Macedon.[58] Polybius is quite complimentary to the Macedonians on their bravery and capacity for hard work, quoting Hesiod on their enjoyment of war.[59] However he has no time for any commanders who, miscalculating the height of walls, find that their scaling ladders are too short.[60] Polybius notes what we would judge to be the Romans' brutality when taking a city but he admired their way of collecting booty for equitable distribution.[61] Polybius is not always complimentary to the Romans, blaming their commanders for naval disasters and criticizing their cupidity and what he took to be poor generalship, stressing the importance of tactics.[62] He thought that Lucius Animus' triumph over Genthius in 168 BCE to be a disgrace, because of the uncouth way he used Greek artists and athletes, and the sack of Corinth in 146 BCE even worse.[63]

Polybius rarely describes how the subunits of the legions were used. The description of the Battle of Zama in 202 BCE. includes details of Roman deployment, explaining how Scipio Africanus planned to reduce the impact of Hannibal's elephants. It appears that Scipio instructed his maniples of *principes* and *hastati* to line up behind each other rather than cover the gaps between the maniples.[64] For modern readers, this has been seen as an insight into how the Romans deployed. However, for Polybius and his audience, the Romans deployment was probably assumed, hence why there is no earlier reference to it. Of course, there is no indication of how big the gaps between the maniples were, they need only to have been little more that the width of an elephant! This measure to reduce the impact of the elephants is also an example of contingent adaptation but one which did not need to be more widely adopted. In Book 18, Polybius describes how, at Kynoskephalai in 197 BCE, Flamininus absorbed his advance guard though gaps between maniples.[65] In the account of the Battle of Zama, we learn that the Romans customarily raised a war cry and hit their shields with their *pila* and spears as they advanced.[66] Again, this incidental detail may have been assumed practice in ancient armies known to Polybius. The conduct of the battle has further detail. The *hastati* were hard pressed by the Carthaginians but leaders of the *principes* brought their units up and attacked the Carthaginians.[67] The implication here is that the *principes* had the freedom to move either through or around the *hastati* but we don't know which or how. We may assume that Polybius' audience did. We are further told that the *hastati* pursued their now-routed enemies but were recalled by a bugle call.[68]

Romans Compared to Macedonians

Polybius fulfils a promise in Book 18, made in a now lost section of Book 6, to compare Roman and Macedonian equipment and formations. Noting that both have been victorious in the eastern and western Mediterranean, he tries to explain why the Romans have always been victorious.[69] Polybius notes that Hannibal used Roman equipment when he could get it and that the famous Pyrrhus had used Italian troops interleaved with his phalanx units but still could not win consistently.[70] Polybius, addressing his Greek audience, states that he provides these comments to avoid contradiction.[71] He begins by emphasizing to his Greek audience what they know: the Macedonian phalanx, sixteen deep, pike-armed with a frontage of 3 Greek feet, roughly 1 metre, was unstoppable frontally.[72] The onset was evidently terrifying, according to Aemilius Paullus who faced the phalanx at Pydna in 168 BCE.[73] Polybius then informs his audience that a Roman soldier, who fights independently, unlike the phalangite, needs twice that frontage with the result that he must face two phalangites but cannot possibly fight through the 4 metre pikes projecting in front of them.[74] Begging the question as to how the Romans can win, Polybius reminds his Greek audience that the phalanx requires level ground without obstacles, otherwise it is disordered. Further, the Romans don't commit their whole forces against a phalanx but use their reserves to attack the phalanx in the flanks and rear. And finally, the phalangite must march into different country and engage in a variety of tasks, yet he is only really suited to one, while the Roman is much more adaptable.[75]

Today, we don't know how Roman soldiers fought. Polybius' contrast between the frontage of Roman and Macedonian infantry has been difficult to interpret but Polybius is clearly saying that the Romans need more space and are not as dependent upon the soldiers either side of them. Instead, they could have fought as a swarm, based around the standards. One meaning of *manipulus*, i.e. maniple, is a group of soldiers who fought around a standard.[76] With this much space, it may have been Roman practice to contract the frontage of a maniple to permit another to replace it in action, during the inevitable lulls in fighting.[77] It has also been argued that Roman infantry did not form up in ranks but rather in a checkerboard pattern, staggered rather than in files 'covered off', in each unit.[78] All of this having been said, we don't know and never will know exactly how the Romans fought although speculation will continue. Armstrong has cleverly suggested that the armies described by Polybius reflected the evolution of the Roman body politic in which different populations came together. What seems difficult to understand militarily may simply be the military manifestation of discrete political/social groups brigaded together; we will never know with any certainty.[79]

To a Greek-speaking reader familiar with the complexities of Hellenistic warfare, Polybius account may have been surprising. The legion integrated soldiers with different equipment into near identical maniples, led by dependable centurions, using cavalry to secure the flanks. Hellenistic practice required a commander to use different kinds of infantry (phalanx, skirmishers, *thureophoroi*), cataphract cavalry, lancers and elephants, in a coordinated plan, relying on subunit commanders.[80] A further surprise was that the Romans did not have a standing army. Hellenistic states, even the cities, had permanently employed soldiers guarding frontiers, in garrisons, as guards for kings or even in police functions.[81] These states had training regimes of different kinds to give men the rudiments of military skill. In time of war but not annually, the population would be called on to supply troops. Those who were military colonists would be obligated to serve.[82] The Romans however were quite different. They raised armies annually. They did not have specialized military commanders but used the elected magistrates. Polybius was particularly concerned that his readers would appreciate one particular aspect of the Romans' capacities: that was their population.[83] His descriptions of both Punic Wars, detailing the sizes of armies and fleets, would have amazed Greek speakers.[84] No Hellenistic state had the capacity of the Romans or the Carthaginians to wage war on this scale.[85] It took little calculation to realize that Polybius' Roman army of four legions with allies would be in the order of 30,000 men and that the Romans were capable of putting a number of armies of this size in the field at the same time using manpower experienced by continual campaigns. By comparison, at the Battle of Raphia in 217 BCE, the two most powerful Macedonian Successor kingdoms of Seleucid Syria and Ptolemaic Egypt, fully stretched their manpower resources to deploy 62,000 and 70,000 men respectively. In Egypt's case this included the unprecedented step of arming native Egyptians. Polybius' broader account of Roman battles showed that the Romans could lose almost that many men at a single battle (Cannae in 216 BCE) and still continue to fight, even in their own country, and win. The defeats of the Macedonian kings Philip V and Perseus, at Cynoschephalai and Pydna respectively, were salutary lessons in the Romans' military capacities, let alone the defeat of Carthage. The famous parade of forces in the games celebrated at Daphne by Antiochos IV in 166 BCE merely demonstrated the comparative weakness of Syria compared to the Romans' capacity to wage war and win.[86]

What Polybius Didn't Say…

As historians, we always seek clarity on all questions, but our sources are not always forthcoming. A particular omission by Polybius, for modern readers, is the lack of a mention of cohorts in his section on the organization of the Romans' armies in Book Six. As we will see in the next chapter, cohorts were the most commonly referred to subunits of legions in Caesar's accounts. Polybius does not refer to them in his description of the Romans' military institutions but he clearly knew of them, even transliterating the term, i.e. 'κοόρτις' (*cohortis*), referring to cohorts in use in Spain in 206 BCE.[87] Livy makes reference to them in Spain too.[88] There are no less than seventeen references to cohorts in Livy for the years 210 to 195 BCE.[89] Archaeological evidence of Roman camps in Spain suggests that some armies encamped by maniples and others by cohorts, with cohorts becoming more a norm in the second half of the century.[90] Armies continued to be organized by maniples with a reference to them in Sallust's account of the war against Jugurtha.[91] Of course, maniples could continue within cohorts given that three maniples of two centuries composed a cohort. The composition of a cohort in terms of the numbers of *hastati*, *principes* and *triarii* is a matter for modern speculation.[92] For Polybius, perhaps cohorts were simply a grouping of maniples which required no particular comment other than an explanation.[93]

Polybius' description of the Romans' military arrangements was composed to answer, for his Greek audience, the question about the reason why the Romans could defeat the previously dominant Macedonians. Criticizing Polybius, as some scholars have done, for not addressing what we might like to know is somewhat unreasonable.[94] Polybius' description suited his literary purpose, emphasizing a contrast he was at pains to draw between Greek and Roman practices. He certainly tried to tell his audience what he felt that they needed to know and understand relative to his purpose in writing.[95] The problem today is that he, quite rightly, assumed that some things would be known, considered to be unimportant or simply not imagined. Within these categories, one could include standing armies of full-time soldiers, the role of the allies, clothing colours, cadenced marching (i.e marching in step), how men were trained, how their food rations were provided and where they sourced their equipment.[96] Much more important than these, we don't really know how Roman battles were fought, although speculation continues.[97] Polybius simply says that the Romans had four types of soldiers in each legion (*velites, hastati, principes* and *triarii*), which fought in separate lines, except for the *velites* who appear to be integrated into each maniple.[98] Zhmodikov has argued that the Romans placed much more reliance on missile exchanges relative to hand-to-hand combat than we have commonly assumed.[99] The length of battles and the view from the ranks have provided a better understanding of how Polybius' legions functioned in battle. The physical exhaustion of combat over hours would demand that soldiers could temporarily disengage.[100] We know that troops from rear lines were, on occasion, redeployed to flanks or rear but generally they appear to have been used to reinforce the lines in front, but we don't know how.[101] It is clear that the Romans were able to replace tired units by fresh ones during a battle.[102] One recent explanation looks to the implications of Polybius' statement that Roman soldiers' equipment meant that they needed twice the space of a Macedonian phalangite.[103] Did this allow units to interpenetrate? It has been suggested, based on a comparison of Hellenistic formations and opposing Roman forces, that there were gaps between maniples even when they were heavily engaged.[104]

Some modern views of what enabled the Romans' military effectiveness institutionally are unaddressed by Polybius. He has also been criticized for what modern scholars have taken to be his overly artificial and simplistic description of how the Romans enlist troops.[105] His description of the *dilectus*, i.e. how soldiers were enlisted, has been questioned, perhaps somewhat unfairly.[106] Magistrates were responsible for raising troops, but they had to receive permission from the Senate. Marcellus, in an effort to have more men, approached the Senate in 214 BCE regarding the use of the survivors of Cannae, but was refused. He was displeased.[107] The Senate, however, gave the consuls permission to draft troops from wherever they wished in 207 BCE.[108] Livy's description is interesting because it reveals the lack of any central command structure to supply additional men. This incident is nothing less than a contingent decision by the Senate because they had no means of otherwise managing the recruitment. This is, of course, consistent with Polybius' description of how the Romans enlisted troops: the consuls were responsible for enlistment.[109] The consul Claudius Nero, in his march to the Metaurus battle in 207 BCE, collected volunteers who put themselves forward, as they were to do for Scipio later. He enlisted those who were of sound body and of military bearing.[110] Micro-management in Italy was not confined to the raising of troops. Prior to one action, Marcellus equipped his infantry with weapons used in naval warfare.[111] The naval weapons were probably long spears, so the ploy is reminiscent both of Camillus' tactics and Flaminius' use of the *hastae* of the *triarii*. While the details of the story are a little unclear, the point of the story for Plutarch is the attention that Marcellus gave even to the equipment of his soldiers. In all cases a general exercised close attention to equipment, even to the extent of changing what we would see as being the usual equipment of a Roman soldier. The uses of spears by both Flaminius and Marcellus are examples of contingent decisions which were not adopted more widely.

There has also been criticism of Polybius' statements about the size of legions.[112] A convincing case can be made for the existence of an elementary bureaucracy needed to administer the Romans' recruitment and service records, another matter outside Polybius interests.[113] Polybius is explicit about Roman manpower but does not address the modern interest in the economy.[114] Our mental maps of combat are modern, not ancient. When modern readers engage with Polybius' description of the Romans' armies, we simply don't understand how, for example, the numerous *velites* were used in battle. Perhaps influenced by Thucydides' rich descriptions of hoplite warfare with its emphasis on hand-to-hand combat at close quarters, Polybius' descriptions of the Romans' armies is a little baffling.[115] Our image of Alexander's conquest of the Persian Empire is similarly based on Arrian's images of close-quarters combat and a decisive charge like Alexander's at Issus in 333 BCE, which was aimed at Darius and broke the fighting spirit of the Persians. Polybius' explanation is that the Romans' soldiers were more adaptable to circumstances, avoiding the rigidity which Macedonian tactics demanded. Modern writers like Hanson have emphasized what has been termed a Western style of battle based on hand-to-hand combat.[116] Such a discourse struggles with a protracted battle of missile exchanges interspersed with eruptions of hand-to-hand combat which we can only imagine. Part of our problem may be our conceptions of what 'light' and 'heavy' infantry were, e.g. skirmishers as opposed to hoplites and phalangites; how can we explain the effectiveness of a Polybian legion when a third appear to be skirmishers?

Polybius does not lay much emphasis on the psychology of battle, although he does note how Scipio Africanus used stratagems to counter Mago's attempt to undermine the confidence of Roman troops at Nova

Carthago in 209 BCE.[117] Polybius probably assumed that his audience would understand the attraction of booty. His near contemporary, the Roman playwright Plautus, offers battle narratives, military metaphors and good wishes to the audience for success in war.[118] For Plautus' audience, *praeda*, i.e. booty, made war attractive.[119] And Polybius says little about mercenaries who were recruited by the Romans but they were a significant part of Hellenistic armies.[120] In the final analysis, we have to accept the editorial decisions made by Polybius and the fact that much of the knowledge that he assumed his audience would have is lost to us today.

Contingent Adaptation in Action: Scipio Africanus, a case study

There appear to have been only minor, if any, permanent changes in Rome's armies in this period but that does not mean that the process of contingent adaptation is not visible. The Second Punic War provides a number of examples of contingent decision-making and micromanagement in which different generals engaged according to the circumstances in which they operated. During this war the Romans deployed armies to Spain for the first time, while, by necessity, they retained the bulk of their forces in Italy. These armies in Spain and Italy were distinctively different for a number of reasons. The forces in Spain were the furthest from Italy that any Roman armies had ever operated. In comparing the Spanish armies with those in Italy and Sicily, there are two key problems with the Spanish armies which must have taxed the Romans. The first was the supply of food and military equipment. The armies in Italy could rely on Italian sources but these could not serve for Spain. The second was the control which the Senate could exercise over the commanders a long way from home. We have no information regarding the second problem, apart from the extraordinary, assumed command of L. Marcius Septimus, who led the Roman armies after the death of the two elder Scipios in 211 BCE; it was a contingent solution to a local situation after the death of two generals.[121]

After a series of earlier commanders and mixed fortunes in Spain, the Romans appointed Publius Cornelius Scipio, the son of one of the previous commanders, to take command of the Roman forces in Spain in 210 BCE. We know this man as Scipio Africanus.[122] The accounts of his activities provide us with a number of examples of contingent decision-making as he responded to the threat to Rome posed by the Carthaginians. The circumstances of the campaign forced contingent decisions to be made regarding the manner in which Scipio equipped, trained and used his forces. It is not entirely clear how the Roman forces in Spain in the Second Punic War were supplied. There would have been an initial allocation of corn, as there was for Scipio's expedition to Africa later.[123] In Rome, some provision was made for the support of the army in Spain. Some resources may have come with the reinforcements sent with Nero and Scipio.[124] The Senate set aside money for clothing the army, though it is unclear whether the money was to be spent in Spain or in Italy.[125] Spain is most likely because it would have been cheaper to transport the money from Italy than the food or equipment. There was food in Spain, which could be purchased or acquired by other means locally. The same could be said for military equipment. In any event, from the references in the ancient texts, it is clear that Scipio felt that he had the responsibility for ensuring that he had the equipment that he needed for his army. On appointment to his command, he probably took other equipment with him but, given the limited capacity of ancient shipping, we should not imagine that he could have taken enough to equip what remained of the Roman forces in Spain. It is likely that Scipio and his successors levied food and equipment from his allies and defeated enemies. There were no more requests to Rome for provisions after 215 BCE.[126] The Roman forces must have relied on local sources of supply rather than a long logistic tail back to Italy. Once Scipio began offensive operations, to a degree, perhaps to a major degree, his forces were maintained by their allies in Spain and by booty.[127] For a modern general, this would be an additional command burden because logistics, though handled by specialists, would still be a concern so far from home. We must also keep in mind that modern armies are much more dependent upon logistics because of the reliance on vast quantities of one-use modern munitions. Significantly, no ancient source saw logistics as a significant part of Scipio's achievement. This is not to say that the ancients were unaware of such things. Hannibal's logistic difficulties in his Italian campaign were well understood.[128] For ancient writers, it was accepted that supply of one's troops was part of the general's responsibility. Scipio's contingent arrangements supplied his army, but within the customary scope of responsibility and contingent adaptation to circumstances of a Roman general.

Livy records that after the storming of Nova Carthago in 209 BCE, Scipio captured war materials and artisans were promised freedom if they worked to produce war materials.[129] Scipio personally supervised the troops, but also the work of the captured artisans in the workshops.[130] It is possible that Scipio's contingent decision to use the men and facilities at Nova Carthago was the origin of the widespread availability of the Spanish swords that so impressed Polybius, even though the Romans had met them earlier.[131] While the elder Scipios might have replaced damaged and lost weapons locally, it was the younger Scipio who first had access to the Spanish arsenals and manufacturing facilities which had equipped the Carthaginian armies. This local change became so universally accepted that Polybius fifty years later described the Spanish sword as the normal equipment of a Roman *legionarius*. This is the earliest example of a contingent decision in response to local circumstances outside of Italy which produced a long-term, though minor, change beyond the life of the particular army and the general responsible for the innovation. Scipio followed up his victory at Nova Carthago by instituting a training regime of manoeuvres, equipment maintenance and repairs interspersed with rest. All of these things were different contingent decisions, the effect of which was to produce a more effective Roman force in Spain.

Scipio was able to make further good use of the products of the captured artisans in Nova Carthago when he beached his ships and equipped his sailors as soldiers.[132] He used the equipment made at Nova Carthago to equip these men. He had previously drafted captives as crews for his ships.¹³³ Scipio was using men in his army who had been drafted for naval service and captives. This is probably not something which Scipio could have done in Italy. Not only was this a breathtakingly contingent measure to increase the size of his army, it also provided the men concerned with opportunities for increased social status and, as such, it may provide a partial explanation for the opposition of the aristocracy to Scipio personally.

Scipio was adventurous in his efforts to win allies, using the powers *imperium* gave him. This included an audacious trip to Africa to win the support of the Numidian prince Syphax in 206 BCE.¹³⁴ The benefits of the training of his troops, which Scipio instituted, are revealed by the hitherto unrecorded capacity of a Roman army to manoeuvre on the battlefield. At the Battle of Baecula in 208 BCE, Scipio avoided a frontal attack on Hasdrubal's strong position by holding the Carthaginian's attention with a diversion by light-armed troops, while sending his heavier infantry around Hasdrubal's flank.¹³⁵ The result was a classic double envelopment, conceptually similar to Hannibal's achievement at Cannae. A similar manoeuvre won Scipio's forces their victory at the Battle of Ilipa three years later. Scipio used a stratagem to trick Hasdrubal into deploying his best troops in the centre. Scipio refused his weaker centre composed of Spanish allies and advanced on both flanks with his Roman troops. Again, both Carthaginian flanks were enveloped.¹³⁶

Scipio's forces were clearly well trained, experienced and confident in their commander. Yet the mutiny during Scipio's serious illness after Ilipa revealed the degree to which this highly trained army was dependent upon Scipio personally. The mutineers' issue appears to have been payment or else booty.¹³⁷ Unlike the armies which defended Italy from the depredations of Hannibal and his allies, Roman armies in Spain were engaged in a different project. For them, the campaigns in Spain were not about defence of hearth and home: they were there for booty. After so long in Spain, the soldiers had grown impatient about their long service away from home. Scipio's solution to this problem was clever. Scipio's resolution came from a levy of allied states to raise more troops and reduce his dependence on disaffected Roman troops.

Rome's Spanish allies had to support Scipio, because without him the Carthaginians would reassert their control and take revenge. The revolt of the Spanish chiefs Mandonius and Indibilis, which collapsed when the rumour of Scipio's death was found to be false, shows the extent to which Roman control was due to Scipio's personal influence. Scipio's development of personal relationships with Spanish leaders was a contingent decision, the more universal practice of which was to lead to the foreign *clientela* which would be a feature of Roman aristocratic relations with subject peoples and states.[138] The resolution of the mutiny concluded with a personal oath to Scipio before the military tribunes.[139] Personal bonds, not institutional ties or arrangements, joined Roman soldiers to their generals in this and later times. Scipio left an army for his successor and settled the rest of his troops in Italica rather than bringing them home.[140] This avoided any threat that bringing his men home would have been to the senatorial status quo, though the reasons probably had more to do with the military situation in Spain, perhaps even a lack of transport. We should not ignore the possibility that the men had developed local connections too, as soldiers did in the Imperial period. Nevertheless, this settlement in Italica was the first Roman colony outside the Italian peninsula. This settlement of former soldiers who had served long in a province was an adaptation to circumstances which was adopted as a practice by later commanders, even emperors.

Scipio returned to Rome in 205 BCE to stand for election to the consulship.[141] He may have held a triumph, though he did not bring his army home. He was elected consul but the Senate was divided about giving him the chance to invade Africa. In the teeth of opposition from Fabius Maximus, Scipio managed to get the province of Sicily with the option of crossing to Africa.[142] This compromise saddled Scipio with no forces except the survivors of Cannae, but these troops were not without both experience and success, having captured Syracuse.[143] Scipio was refused permission by the Senate to levy troops. He was still able to accept 7,000 volunteers, probably because that could not be refused.[144] While he could not levy additional troops, Scipio was permitted to take to Africa troops assigned to Sicily. His response was a contingent decision in light of the circumstances. He used in part the survivors of Cannae. He selected men who had experience in the siege of Syracuse.[145] He inspected these troops and took all who were fit for service and made up deficiencies in numbers from the volunteers.[146] In Sicily, he organized the volunteers into centuries and found a way to turn 300 of them into cavalry at the Sicilians' expense – another example of an increase in social status for the Roman infantry concerned. Having manned the legions, he drilled them.[147] Cato evidently found fault with the manner in which Scipio prepared his men for the campaign in Africa. Cato's objections stressed how Scipio spent his financial resources. It is not entirely clear from Plutarch's text, but the financial resources may have been Scipio's private funds.[148] In the account of the investigation of what we know as 'the Pleminius affair', Livy describes a praetor and the *legati*, i.e. representatives, inspecting a display by Scipio's land and naval forces and touring the arsenals and store houses in Syracuse. They were impressed with Scipio's preparations for the assault on Carthage.[149] Scipio had paid close attention to what he needed to undertake the invasion of Africa. To preserve the grain supplies from Italy with which he had been provided, Scipio billeted his troops on luckless Sicilians.[150] Again he had found a way to equip his army without the support of significant material resources from Italy. Scipio's attention to detail in the sea journey to Africa extended even to details of the lights to be shown by the ships.[151]

Our only measure of Scipio's African army's efficiency is its performance in the field. Scipio's problem once he landed in Africa was to bring his new army to something approaching the peak of his old Spanish army. He evidently decided to use the same approach to training and tactics that he had found to be successful in Spain, though much of the training had already been conducted in Sicily. The surprise attacks on the camps of Syphax and Hasdrubal would have been a good preparation for a larger battle, instilling confidence in the troops and providing them with booty.[152] The battle in the Great Plain soon after was the first large-scale action in which Scipio's African army fought. The destruction of the Carthaginian cavalry on both wings left the Celtiberian mercenaries completely exposed. Livy's text indicates that they were surrounded.[153] Polybius specifically says that the Celtiberians were surrounded by the *triarii* and the *principes*.[154] Given the usual Roman deployment of these troops behind the *hastati*, this meant that the *triarii* and the *principes* must have been manoeuvred around the flanks of the Celtiberians to envelop them.[155] Scipio probably executed a double envelopment also at Zama, though Livy's confused account is not lent much credibility by the reference to a Macedonian legion and superior Roman war cries. In any event, it would appear that Scipio deployed his forces differently from conventional Roman practice, in that he left gaps between his maniples of *hastati* to allow the Carthaginian elephants through. These gaps were covered by *velites*, who had orders to let the beasts through.[156] Appian has a similar account with the addition that the *velites* were given iron-shod poles and were told, sensibly, to avoid the elephants frontally. The cavalry, which was deployed on the flanks by Scipio were assigned, as support, men on foot armed with javelins.[157] After the Carthaginian cavalry were routed on each flank and the mercenaries and inferior Carthaginian troops in the first two lines were eliminated, Scipio appears to have redeployed his infantry. Livy describes Scipio rallying his *hastati* and then moving the *principes* and *triarii* to the flanks before the final engagement with Hannibal's veterans.[158] Polybius too describes Scipio redeploying the triarii and the *principes* on both flanks and in line with the *hastati*. This produced another envelopment, again on both flanks, though Livy and Polybius both claim that the Roman cavalry, returning from pursuit, did the greatest execution, perhaps attacking the rear of Hannibal's veterans and preventing their escape.

Scipio Africanus had developed a way of winning battles which stressed the use of stratagems and the redeployment of troops from the rear lines to attack the enemy's flanks. Africanus was not the only Roman general of this period to do something like this. At the Battle of the Metaurus in 207 BCE, Claudius Nero moved troops from the rear line of the right wing around the back of his army to attack on the left and so take the Carthaginians in the flank.[159] Scipio was, however, the only Roman to carry out double envelopments. It is notable that Polybius' account of the superiority of Roman military organization and equipment to that of the Macedonians refers to the way the Romans responded to the irresistible charge of the phalanx by attacking its vulnerable flanks. Livy's account of the Battle of Kynoskephalai, where an unnamed tribune directed twenty maniples from the victorious Roman right wing to attack the Macedonian right, shows the value of this approach.[160] This counter-attack is typical of the flexibility which Polybius believed to be inherent in the Roman organization as opposed to the Macedonian.[161]

There is nothing in any ancient source which accords Scipio or anyone else the development of this flexibility, yet it remains that we do not see it exploited in this manner before the Second Punic War. The simplest way to explain it is that Scipio showed others the possibilities which can arise when an army is together long enough to develop the level of competence to perform complex evolutions on the battlefield as Scipio contingently adapted to the circumstances. The tactics of Scipio were the product of a combination of a series of contingent decisions which permitted the Roman armies to adapt to those of the Carthaginians. These adaptations, when combined with experienced soldiers and officers, enabled generals to undertake more sophisticated manoeuvres. We should note that the army with which Flamininus defeated Philip V at Kynoskephalai in 197 BCE, by attacking the flank of the phalanx, contained veterans of the Punic War and Scipio's veterans specifically.[162] There are no reforms here, but instead we see the exploitation of the advantages inherent in Roman organization explained by Polybius, the full benefit of which could only be realized with well-drilled experienced troops led by experienced commanders. Further, we should note that there appears to have been no effort to pass on the knowledge gained by Scipio's and Nero's successes. Apart from any other considerations, there was no mechanism and no body to perform the task, unlike those which exist to support modern armies.

Chapter 7

Rome's Armies in Caesar's Time

Introduction

After Polybius, the works of Julius Caesar and his continuators provide the largest body of evidence on Rome's armies down to the death of Augustus. The Roman armies commanded by Julius Caesar have been researched for many years. Judson's 1888 book, using the latest German research is one of the earliest in English and Dodge's 1894 contribution was supported by his in-person inspection of Caesar's Gallic battlefields.[1] Even a casual reading of the works attributed to Julius Caesar and his near contemporaries, with an eye to the Romans' military institutions, will suggest that Rome's armies appear to be different to those described by Polybius'. The principal differences between Polybius' and Caesar's armies are the use of the cohort as the tactical unit of manoeuvre, the disappearance of Roman citizen light infantry (*velites*) and of Roman citizen cavalry.

The replacement of the maniple by the cohort is the major component of the so-called Marian army reforms, referred to in Chapter 3, with one or more of the other changes included in them by various scholars.[2] While 'reform' has been shown to be an inappropriate term to use for the Romans, a consideration of Marius' role in change in the armies is still required. It has also been asserted, falsely, that 'the flexible cohort structure seems to have been made possible by the professionalism of the army', although, as we have seen, 'professionalism' is an inappropriate term for Rome's soldiers in this period.[3] The appearance of the troops appears to have changed, although our evidence of the appearance of Roman soldiers in this period is limited by the few reliefs and slender archaeological evidence. The Aemilius Paullus monument and the object known as the Altar of Domitius Ahenobarbus provide what is perhaps an idealized image of Roman soldiers in monuments glorifying aristocrats (see Figure 12).[4] However, there are textual references which provide a different perspective.

C. Marius

The different organization of Caesar's army compared to the army described by Polybius has led to the view that there had to be an army reform and a reformer between 130 and 59 BCE. The reformer was identified to be Gaius Marius and these reforms are commonly known as the Marian Reforms.[5] The problematic nature of the use of the term 'reform' has already been discussed as has the use of the term 'professional' to describe Roman soldiers in this period. Apart from the unsuitability of 'reform' and 'professionalism' in reference to the Romans, the evidence for the changes accorded to C. Marius have been thoroughly reviewed by François Cadiou, who believes that the author of the idea of a Marian army reform was Fustel de Coulanges in the nineteenth century and that his construction has been accepted too readily ever since.[6] Cadiou's conclusion from his review of the evidence is that Marius is only responsible for the innovation of accepting volunteers from the poorest census class, nothing more.[7] The Romans of a later time certainly saw in Marius' acceptance of volunteers the beginning of political changes which are often seen as the destruction of the Roman republic but this is not army reform.[8] In the decades that followed, there were momentous political events, the Social War with the Italian allies and two civil wars before Augustus became *princeps*, i.e. leading man. Marius did not make permanent changes to Rome's armies and there is no evidence that he did.

Scholars have struggled with the idea that Marius reformed the army for a considerable period of time. Delbrück, rejecting Mommsen's suggestion that Rutilius Rufus was responsible, believed that cohort tactics arose from the Marian Army Reforms.[9] Kromayer and Veith acknowledged the contentious nature of what they termed 'the "so-called" Marian reform' (*'die sogenannte marianische Reform'*).[10] In spite of their recognition of the concept as 'an arbitrary construction' ('*eine willkürliche Konstruktion*'), they argued that the changes are of such a unity as to require a single author which they believed to be Marius. Parker believed that Marius is most likely to have been the one to replace the maniple with the cohort as the basic formation of the Roman infantry, though he did acknowledge that the attribution of this reform to Marius is not made on the basis of any direct evidence.[11] Webster believed that the change to the cohort was an unintended outcome of the recruitment of the *capite censi*, the lowest, poorest census class, making the reform an indirect effect of a change in recruitment.[12] Keppie cautiously attributes to Marius the change to cohorts. He, agreeing with Webster, sees a good argument to be that since the state supplied the men with their equipment, the previous distinctions based upon wealth would have been unnecessary.[13] Bell begins, as others have, with an acknowledgement that no ancient text ascribes reform to Marius.[14] Bell is particularly interested in the conflicting evidence for the use of both cohorts and maniples as units of organization before and after Marius. His explanation is ingenious. Maniples and cohorts were both used before and after Marius because maniples were suitable for more 'civilized' opponents like the Macedonians, but cohorts were better against 'barbarians'.[15] Bell, acknowledging the evidentiary problems, believes that it is difficult to assign any change to any one man, but he gives Lucullus responsibility for the final disappearance of the *velites*.[16] Bell proposes that the adoption of the cohort and the disappearance of the *velites* were gradual and complete by Caesar's time.[17] Kertész has pointed out that there are references to the use of cohorts that pre-date Marius' time.[18] Taylor does not believe that the changes in the armies in the late Republic were due to Marius and that the use of cohorts was due to the formations of the Italian *socii*, allies.[19]

Marius is recorded as enlisting *capite censi*.[20] The enlistment of such men may not be unprecedented if Crawford understands Ennius' reference to *proletarii* correctly to refer to men below the minimum property qualification.[21] Marius' acceptance of *capite censi* for enlistment into the legions has been represented in modern times as marking the introduction or development of professionalism, in spite of other claims for Scipio Africanus' responsibility. We have already seen that 'professionalism' is an inappropriate and unwarranted concept to apply to Rome's armies but its attraction to some modern scholars was what they saw as a connection to the creation of a 'standing army', i.e. troops enlisted full-time in peace and war. Much of the discussion in these terms has reflected recent contemporary political views and assumptions based on modern experience. Cary described the army before Marius as a conscript militia while the army after Marius is styled as a standing force of professional warriors.[22] Cary, applying a modern term, described the *capite censi* as 'urban proletariats', reflecting an anachronistic view of the poor in Rome.[23] Scullard saw Marius in much the same light as Cary. Parker believed the significance of this change was that the institution of a professional army was officially recognized.[24] Parker's footnote to this referred the reader to Delbrück. Smith claims that 'what was in effect a standing army' existed prior to Augustus.[25] This led him to Marius as the origin of these armies. Smith's claim is based upon the anachronistic view that such soldiers had chosen a 'military career'.[26] Gabba claims that the armies were now professional.[27] Harmand also believed that Marius replaced citizen troops with permanent professionals.[28] Webster saw the need for professionalism and describes the *capite censi* as the proletariat.[29] Smith's language suggests that he was aware that he was applying a modern construction to the ancient situation: 'it became possible to create what was in effect a standing army'. This suggests that he was aware that the Romans' arrangements were not consistent with the modern meaning of 'standing army'.

Little is gained by describing these forces as 'standing armies' even with a qualification. The term normally refers to armies which exist regardless of whether the nation was at peace or war. Smith assumes that the Romans always had troops in Asia though he admits that the evidence is lacking.[30] His argument is based upon the existence of the Romans' dangerous neighbours, yet these neighbours lacked standing armies of their own as far as we know. Looking at the provinces, there is no evidence for Roman forces in Bithynia-Pontus throughout the period. He assumes that the Romans always had forces in Cisalpine Gaul.[31] There were certainly forces in Transalpine Gaul, the Spains and Macedonia, but there is no evidence that these forces were there for any other purpose than to engage in the Romans' endemic expansionary wars and a comparatively small role supporting tax collection. There is no evidence that standing armies existed whose purpose could be defensive as well as offensive.[32] Syria could be a different case except that, again, all Smith offers is a series of commands. In all these cases, a series of commands and commanders do not of themselves support the existence of standing armies. All they show is that the Romans assigned the provinces to men who had forces at their disposal. The most obvious explanation is that these *provinciae* gave the commanders the opportunity to show their *virtus* as Cicero did during his command in Cilicia. The forces were not defensive garrisons, that is forces to secure a province in peacetime, any more than the Roman forces sent to Spain in the second century were. The use of the term 'standing army' for Roman forces in the first century BCE is anachronistic, inaccurate and misleading, and Marius was not responsible. It is much simpler to see Marius' enlistment of *capite censi* as a response to the Senate's attempt to frustrate his *supplementum* to raise more troops rather than any intention to make an innovation.[33] The claim of professionalism in these armies has been directly rejected by Gabba and Brunt.[34] Brunt has shown that the application of the concept of professionalism to armies of this period lacks evidentiary support.

The most important reason for the abandonment of the claim that Marius replaced the maniple with the cohort as the unit of tactical manoeuvre in the legions, the major component of these reforms, is that the use of cohorts pre-dates him. Livy refers to cohorts among allied troops in 294.[35] Livy refers to cohorts in Spain prior to Marius.[36] He mentions a *cohors Romana* in Spain in 211 BCE and at the Metaurus in Italy in 207 BCE.[37] Polybius specifically mentions 'Roman cohorts' (Ρωμαίοις χόρτις), at the Battle of Ilipa in 206 BCE.[38] While Livy and later authors like Appian may be anachronistic and somewhat untrustworthy on military terminology, Polybius' testimony cannot be ignored, though

1. Niccolò di Bernardo dei Machiavelli (1469–1527) by Santi di Tito. (*Wikimedia Commons*)

2. Raimondo Montecuccoli (1609–1680) by Elias Grießler (1622–1682). (*Wikimedia Commons*)

3. Jean Charles, Chevalier de Folard (1669–1752). (*Millearia via Wikimedia Commons/CC BY 3.0 DEED*)

4. Maurice de Saxe, i.e. Hermann Moritz von Sachsen (1696–1750) by Maurice Quentin de La Tour. (*Wikimedia Commons*)

5. Antoine-Henri Jomini (1779–1869) by workshop of George Dawe. (*Wikimedia Commons*)

6. Hans Gottlieb Leopold Delbrück (1848–1929). (*Wikimedia Commons*)

7. Certosa Situla, from Etruscan antiquities in the Archaeological Museum of Bologna, with illustrations of figures with different equipment in the top row. (*Wikimedia Commons*)

8. The front cover of von François's *Tannenberg*. (*The Internet Archive*)

9. Parade of the 'Prussian' 1st Foot Guard Regiment in the Lustgarten, 9 February 1894 (Potsdam) by Carl Röchling. (*Wikimedia Commons*)

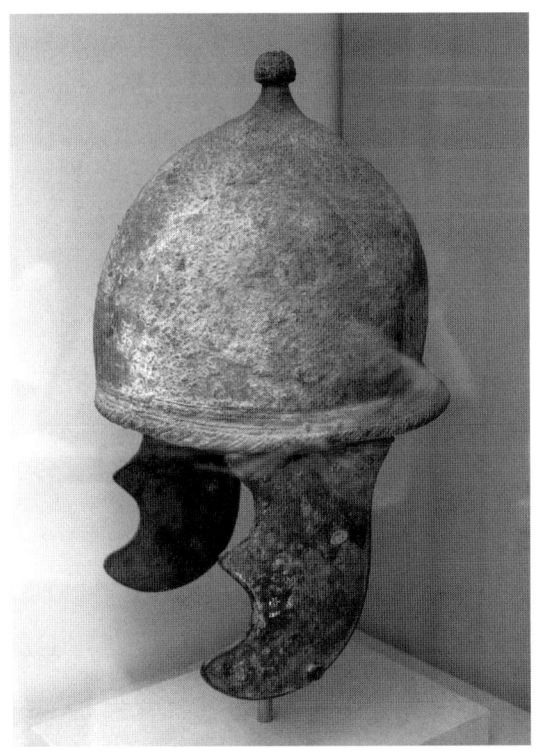

10. South Italian bronze helmet from the third century BCE. (*The British Museum*)

11. Section of the Pydna Monument showing two helmets. (*Colin Whiting via Wikimedia Commons/CC BY-SA 4.0 DEED*)

12. Roman soldiers on the monument known as the Altar of Domitius Ahenobarbus. (*Wikimedia Commons*)

13. Commemorative stele of Centurion M. Caelius who died in Varus' forces in Germany. (*Wikimedia Commons*)

 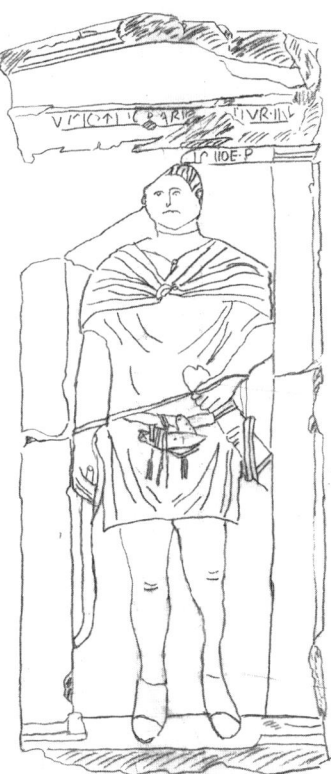

14. Sketch of a monument from Urso, Spain, showing Roman soldiers, Republican period from the shield shape.

15. Sketch of the funeral stele of Centurion Minucius, Legio Martia in Padova Musei Civici.

16. Modern re-enactors in France, at Vienne, in 2015.

its significance is uncertain. He does not refer to the cohorts when he describes to his readers how the legion was organized.[39] Parker acknowledges Polybius' text but dismisses it as referring to a temporary expedient; perhaps it was.[40] In any event, there is no evidence that Marius institutionally or even temporarily changed Roman infantry organization from maniples to cohorts.

The disappearance of the *velites* is commonly listed also as a feature of the Marian army reforms.[41] Marius cannot be responsible for the disappearance of the *velites* because they appear in later Roman armies, though there is no sign of them by Caesar's time. There is a reference to *velites* in Sallust's account of the Jugurthine War, 112–106 BCE during which Marius served; Metellus has slingers and archers, presumably provided by allies, as well as *velites*.[42] Most significantly, the latest reference to *velites* refers to their deployment by Sulla at the battle of Orchomenos in 85 BCE, well after the time in which Marius is claimed by modern authors to have reformed them out of existence.[43] As to their eventual fate, Bell assigns the disappearance of the *velites* to Lucullus.[44] This claim is at best a guess and unnecessary. The actual reason was probably that Roman *velites* used javelins and were outranged by slingers and archers.[45] Over time, Roman magistrates, adapting to circumstances, probably ceased enlisting them, preferring heavier-armed troops and slingers and archers provide by *socii*.

It is clear from both Livy and Polybius that wealthier Romans served as cavalry during the second century but Marius is claimed by some scholars to have removed them under his reforms.[46] Citizen cavalry were probably difficult to recruit for some time before Marius. Polybius comments unfavourably on the reluctance of young aristocrats to serve in Spain in the 150s BCE.[47] These would surely have mostly served as cavalry, not infantry. However, there is evidence that Roman citizen cavalry was available in Marius' time and later. In 109 BCE Sulla brought a large force of cavalry from Latium and the Italian allies to Numidia to fight Jugurtha.[48] Marius himself had a cavalry bodyguard in Numidia shortly after and it appears to have been composed of Roman troopers.[49] With Italy under threat, Roman citizen cavalry served in the Cimbric War when Marius commanded legions. In c.101 BCE, the son of the *Princeps Senatus* (leader of the Senate), M. Aemilius Scaurus, was a member of a unit of Roman cavalry that fled to Rome following a Cimbric attack at the River Athesis.[50] The deployment of such cavalry for the defence of northern Italy would be consistent with both a substantial threat to Italy and the need to raise local cavalry. Pompey himself surrendered the 'public horse' (*equus publicus*) given to him as a citizen who could serve as a cavalryman in 72 BCE, after having rendered his required years of service, though this does not necessarily mean that he served as a cavalryman.[51] However Pompey's army to oppose Caesar in the Civil War contained 7,000 cavalry, whom Plutarch describes as the flower of Rome and Italy, highborn, rich and spirited, but Plutarch confusingly also implies they were drawn from many different foreign states.[52] Caesar's account of his troops' resistance to Pompey's cavalry attack does not refer to them as Romans or as anything else for that matter.[53] If they were non-Roman, he would have had little reason not to be specific. If they were Roman, he would have had no reason to gloat.[54] There is no good reason to doubt that this cavalry was Roman, reflecting a civil war situation. This is not the last we hear of Italian cavalry; Octavia sent Italian cavalry to Antony in the East.[55] These references are consistent with the reference in the *Tabula Heracleensis*, dating from Caesar's time, to service as a cavalryman in a legion: 'unless those who served as a cavalryman for three years or as an infantryman for six' (*nisei quei eorum stipendia/ equo in legione [tria] aut pedestria in legione [sex] fecerit*).[56] If the Roman cavalry had been abolished by Marius, what would have been the point of this regulation? Marius simply could not have been responsible for the disappearance of citizen legionary cavalry since they existed after him.[57]

There is no ancient writer who assigned to Marius anything like a substantial tactical change in the equipment, organization or tactics of the Roman army, though he is credited with some innovations for the troops he raised or commanded, except for the very late source, John the Lydian, 600 years after Marius.[58] This point is critical. Regardless of what he might have done for his own troops, there is little to support claims of a wider institutional change. Marius is said to have changed the design of the *pilum* by replacing one of the metal pins, which fixed the head to the shaft, with a wooden one, and to have introduced eagle standards; these are almost trivial changes.[59] They cannot be construed as reforms in any terms. The question of Marius' responsibility for the introduction of professional soldiers into the Roman army is to a great degree a function of definitions and paradigmatic thinking, as we have seen earlier. Sallust describes how the Senate, to frustrate Marius, voted a *supplementum*, believing that the people would not volunteer because of the unpopularity of military service.[60] Marius in response enlisted any men who presented themselves, i.e. volunteers, ignoring the census classes. Sallust notes the different opinions as to Marius' motives. Some believed that he was short of suitable men, while others saw it as an attempt to gain favour with the populace.[61] Smith, whose focus is on the impact of the introduction of the *capite censi* into the Romans' armies, acknowledged that Marius' action in accepting the *capite censi* was a response to a long-standing problem of recruitment.[62] Marius' acceptance of *capite censi* may have been a precedent, but it was not a legislative change. Other generals who followed may have accepted his actions as precedent, but our poor knowledge of the period after Marius' victory over the Cimbri and Teutones means that it is impossible to know. Whatever happened later after the Social War in relation to military service occurred twenty years after Marius' emergency measure. It would be more sensible to look at the effect of the Social War and the role of contemporary circumstances. In any event, Roman soldiers were still recruited based on the census after Marius' death.[63] Marius' actions are more easily explained as contingent adaptation to the circumstances which he faced.

Caesar's Armies

If we discard Marius' reforms, we have a significant challenge to understanding Rome's armies after the death of Polybius, possibly as late as 120 BCE.[64] This period is marked by the Social War between

Rome's Italian allies. Allied troops fighting as part of Roman armies are mentioned in Polybius' time. After the social war, when many Italian allied states rose in revolt against the Romans, they are rarely heard of later.[65] But peace was not the result of the Social War. It was followed by Cataline's revolt in 63 BCE and by the civil wars between Caesar and his enemies in 49–45 BCE and the later wars of the triumvirs after Caesar's death from 42–31 BCE. However, these conflicts will not be examined in detail, although attempts have been made to trace legionary formations from the triumviral wars as precursors of imperial legions.[66] Our aim is to look at the armies to explain their military operation, as Polybius did. The focus instead will be on following Polybius' model, describing equipment, organization and tactics. After Polybius, for detail on Roman armies, we must wait until Caesar and his continuators began releasing his self-serving commentaries on his campaigns in Gaul, Greece, Spain and North Africa between 60 and 44 BCE, although the authorship of books on the wars in Spain, Africa and Alexandria is uncertain.[67] Sallust, a contemporary of Caesar, provides useful detail in his two surviving works on the Jugurthine and Catalinarian Wars, and Plutarch's biographies, written much later in the imperial period, are valuable. Appian, who wrote later still but clearly had access to sources now lost to us, is useful too. Given the sheer volume of Caesar's work, we may begin with his representation of the armies as a snapshot view which can be enlightened with material from other authors; we will never be able to build a connected narrative from Polybius to Caesar and beyond; the sources are lacking. There are, however, numerous examples of generals' contingent adaptation to the circumstances in which they and their armies operated.

We may begin by considering how armies were raised. Polybius describes how armies were recruited in Rome. As discussed above, it is well-known that Marius, in the emergency circumstances of the threat of the Cimbri and Teutones, accepted volunteers from the lowest census class, the *capite censi*, but there is no evidence that this was anything more than an emergency measure and a contingent adaptation to circumstances. In the later wars, we simply don't know how soldiers were enlisted in general but the *dilectus*, selection as described by Polybius, probably continued to be the means.[68] We can suspect that there was a tension between *mos maiorum* ('the customs of the ancestors'), the pressures of social and civil war and the costs of equipping men who owned little or no equipment. Politically, generals with money had the capacity to raise armies either from their own resources or from money lenders.[69] The size of legions in this period is unclear. A veteran unit could get down to 1,000 men. An analysis of Caesar's armies in Gaul suggests an average size of 4,000 men per legion.[70] Most legions in this period seem to range between 3,000 and 6,000.[71] We have no statement equivalent to Polybius' information on the size of legions in this period. Polybius provides a range of 4,200 to 5,000 men although this should be seen as somewhat theoretical given the exigencies of campaigning. The nominal size of legions in the first century BCE may have been as large as 6,000.[72] Brunt's survey of the ancient evidence suggests that the average size of legions in the period is closer to 4,000.

The period of the civil wars, with variations in strength between newly raised and veteran legions, needs to be considered, as does the quality of veteran troops.[73] There is a description in Appian of a battle between the troops of M. Antonius and those of 'Carsuleius' (probably Decimus Carfulenus) and Pansa near Mutina in 43 BCE. We are told that, because both forces were veteran troops, no battle cries were used because neither would be scared. The battle is described as a soldiers' fight where they battled in silence and, when tired, drew apart to rejoin the battle later.[74] Incidentally, this comment tells us something about the individual combat style of the Romans. Appian comments on the benefits of experience which he believes to be more important than ethnicity; both armies were Italian.[75] Aulus Hirtius emphasizes the importance of experience when he contrasts veteran legions with Legio VIII which had only seen eight years' service.[76] Evidently, it took a considerable time to attain the expertise displayed by the soldiers described by Appian. One of Caesar's veteran centurions had been in service for thirty-six years.[77] Veterans could also be hard to control.[78] Yet veteran troops were valuable and could be trusted in difficult situations whereas recruits could not be relied upon.[79] Antonius was prepared to pay each of the veteran Legio XXVIII 500 denarii if they fought for him.[80] Perhaps one reason for the sharp distinction between veteran and newly raised troops is found in Valerius Maximus, where he claims that the consul of 105 BCE, Rutilius Rufus, was the first general to give his soldiers arms training; he used gladiatorial instructors.[81] This implies that Roman soldiers were expected to 'learn on the job'. It is hardly surprising then that veterans were so valuable, they had learnt and survived. Recalling that Polybius detailed the length of service expected of Roman citizens, which implied some kind of nominal list, Appian tells us that the Romans kept records of soldiers' characters.[82] It is unclear where these records were kept and for how long. There is no known central repository of such information. It is more probable that any such record would be maintained at unit level rather than centrally. For the imperial period, when Appian wrote, some papyrus records have been recovered from Egypt, although they do not record soldiers' characters.

Not all legions may have been raised from Romans.[83] Caesar raised Legio V Alaudae in Transpadane Gaul from locals and later enfranchised them.[84] Pompey's troops may not have all been Roman; there is the mention of a *legio vernacula*, i.e. native legion, Spanish troops trained and equipped to fight in the Roman way. A third of Caesar's opponents' army in Spain had been raised there.[85] This may not be the only such unit.[86] Cassius Longinus raised such a unit for Caesar in Spain.[87] But we should be cautious in assuming that such units were not composed of some Romans. Earlier, Cicero raised two legions of Romans from his province of Cilicia in 51 BCE.[88] Civil war necessities had Brutus train Macedonians, for whom he had great respect as soldiers.[89] Deiotarus, king of Galatia, an ally of Caesar, trained two legions in the Roman fashion.[90] One such legion later became the imperial Legio XXII Deiotariana.[91] While the Social War had ended 40 years earlier, Caesar refers to Curio's army as having cohorts composed of men from Italian peoples, Marrucini and a Pelignian, described as from the weakest units in Curio's army.[92] This suggests that the distinctions between Romans and other Italians had not disappeared entirely. Of greater importance to this study is the willingness of commanders to respond to their circumstances with contingent adaptation which led them to enlist non-Romans and, in at least the cases of Legio V Alaudae and Legio XXII Deiotariana, the adaptation led to a permanent situation which enfranchised the troops.

A particular feature of Rome's armies after Marius' victories over the Cimbri and Teutones is the need to fight other Romans or Italians. Whereas Polybius' armies never fought other Romans and only fought some Italians during the Second Punic War, Social and Civil War armies in this period often did. The result could be brutal when Roman troops looted other Romans as vigorously as they looted foreigners.[93] The Roman discipline, admired by Polybius was much less evident in the circumstances of civil war when soldiers were willing to betray generals for survival.[94] However, there are examples from this period of Roman soldiers showing discipline and not stopping to plunder, relying on a just distribution.[95] There are also examples of generals, looking for popularity, allowing men to plunder and other examples of generals who could not prevent it.[96] Even Caesar's troops were affected.[97] In relation to discipline, a particular feature of the armies of this period is the assertiveness of soldiers, sometimes insisting that their generals fight.[98] To a degree, the different picture of Roman soldiers in this period is a product of the literary purposes of the authors, some of whom, like Sallust and Appian, are highly critical of the civil wars, and others, like Caesar, who are at pains to present themselves in the best light for their own political purposes. Appian claims that the indiscipline of Roman soldiers during the civil wars was due to the way in which they were enlisted, which was not consistent with Roman custom.[99] One scholar has identified twenty-four mutinies reported during periods of civil conflict between the outbreak of the Social War and the Battle of Actium and a further eight not connected with civil wars.[100] There were, in addition, even more cases of mass desertion.[101]

Comparing Polybius' description of Rome's armies and depictions of Roman soldiers in this period, a marked difference is that the *velites*, the legionary light infantry, are not mentioned after 80 BCE. Sallust describes them in Metellus' operations against Jugurtha in 109 BCE although slingers and archers operating between the maniples are also described.[102] The disappearance of the *velites* is quite significant as they constituted almost a third of a Polybian legion. A Polybian legion had only 3,000 heavily armed infantry with the balance of 4,200 infantry being lighter-armed *velites*. Legions in Caesar's time lacked *velites* but would have had proportionally more heavily armed men. As noted in the preceding chapter, *velites* could fight hand to hand as well as skirmish with their javelins. Further, they were integrated into the maniples and probably not deployed separately as there are no additional centurions beyond the maniples to direct them - Polybius is clear on this. We may only guess why they disappeared but the reason may simply have been that troops could be equipped with the panoply of the better-protected heavy infantry and, if the importance of the use of missiles by troops with *pila* is correct, providing the former *velites* with better defensive armour would not have compromised missile fire to any great extent.[103] A law associated with C. Gracchus required the state to provide soldiers' clothing but the details are quite uncertain and we don't know if equipment was included.[104] Non-Italian allies could provide archers and slingers who outranged javelin-equipped missile troops like the Polybian *velites*.[105] Aulus Hirtius, the author of Book Eight of Caesar's *De Bello Gallico*, and the probable author of *De Bello Alexandrino*, refers to *auxilia levis armaturae*, i.e., light-armed allies.[106] These light armed troops may have been the replacements for the *velites*, although we should not forget that Cicero told the Senate that *sociorum auxilia* ('the assistance of the allies') was feeble because of the injustice of Roman provincial administration.[107] Another group which appear to fulfil the functions previously performed by the *velites* are described as *expediti*, i.e. 'unencumbered' or 'ready'.[108] Caesar used the term to describe whole legions moving quickly, presumably not encumbered by baggage.[109] These troops were drawn, on at least one occasion, from recalled veterans, (*evocati*) and archers and slingers.[110] There are also references to troops described as *antesignarii*, 'men in front of the standards'.[111] These troops appear to be more lightly burdened than other troops but it is unclear how they were equipped.[112]

Infantry, including Italian allied troops, in this period are organized by cohorts. Scholars have speculated that warfare in Spain required a more convenient grouping between maniple and legion but this is by no means certain.[113] Others have suggested that the use of cohorts reflected professionalization, providing greater tactical 'flexibility' which Caesar utilized.[114] Discarding professionalization, the issue of flexibility remains as a matter of perception. Maniples still existed.[115] The term is found in Latin of Polybius' time in the work of Plautus and continued in use.[116] Caesar refers to members of maniples as *'manipulares'* as does his contemporary Cicero. *'Manipularis'*, a member of a maniple, may have been a common term for a soldier, something like 'squaddie', 'digger' or 'GI' for British, Australian and US soldiers respectively. The word *'cohors'*, i.e. 'cohort', implies an aggregation or a grouping just as maniple implies a handful (from *'manus'*, hand). There is much evidence, particularly in Livy, that cohorts were used for organization during Polybius' time.[117] Confusingly, even later in the Imperial period, centurions are identified by their maniples, even using Polybius' divisions into *hastati*, *principes* and *triarii*.[118] We may safely assume that the armies in Caesar's time had six maniples in a cohort as Polybius describes. Nevertheless, Caesar's use of terminology does not provide great certainty.[119] The lack of consistency in the use of the terms also complicates our understanding of another change. At some point between Polybius and Caesar, the *triarii* ceased to use *hastae*, spears, replacing them with *pila* (Roman heavy javelins). There has been speculation that *hastae* were abandoned because they were unsuitable to a legionary organization based on cohorts. This uniformity may have made their tactical use similar to that of other cohorts.[120] Further, during the Imperial period, the cohorts of the *triarii* seem to have been the same size as other cohorts, not half the size as Polybius describes, but we don't know when this change occurred.

The Romans continued to use cavalry but they were not always provided by Roman citizens. Caesar used Gallic and German cavalry. During the civil war between Caesar and Pompey, Pompey fielded 7,000 at Pharsalus and 3,000 died at Munda.[121] While, as we have seen earlier, Marius cannot have abolished Roman citizen cavalry, it remains that

Caesar's legions in Gaul lacked citizen cavalry. Having accepted that Marius could not have reformed citizen cavalry out of existence, there have been attempts to explain this as a result of the Social War.[122] This is not convincing. The solution may lie in how legions were raised. The Polybian description of the *dilectus* is the only one we have. Polybius tells us that 300 cavalry were assigned by the censors before the infantry were selected, a change from an earlier practice when the cavalry were selected after.[123] When the army assembled, the allies supplied the same number of infantry as the Romans but three times the number of cavalry.[124] As the Romans had always used more allied cavalry, provided as part of the *auxilia sociorum*, the assistance of the allies, we should not be surprised if after the Social War, when most of the Italians were Roman citizens, the Romans continued to rely disproportionately on allied cavalry, not Roman citizens who now included the previously allied Italians. Nevertheless, the decision as to who decided on the number of citizen cavalry to be assigned to a legion must still have been based on census lists and the consul who enrolled the troops. The lack of censors between 86 and 70 BCE and then the election of censors in 65 and 64 may suggest that Roman cavalry lists were simply out of date at the time when Caesar's army for Gaul was raised. Caesar may have decided that he did not want to raise a large number of citizen cavalry or he was denied them, although *contubernales* (generals' bodyguards and aides) did serve Caesar mounted.[125] This would be simply a contingent response to circumstance.

The equipment of Roman soldiers in Caesar's time appears to have been different to that described by Polybius who tells us that Roman *hastati* and *principes* carried two different kinds of *pila*, the Roman heavy javelin. These javelins continued to be used and archaeological evidence shows a variety of designs with different heads, means of attachment to the shafts and, we may guess, different shaft lengths, corresponding perhaps to Polybius differentiation into light and heavy.[126] One of the changes which Marius is claimed to have made was to fix the javelin heads to the wooden shafts by a pin so that the *pila* could not be thrown back.[127] This contingent decision to change the construction of some *pila* was not continued later and its effectiveness has been questioned.[128] In an account of a surprise attack, Caesar tells us that the *pilum* could not evidently be used effectively if the enemy attacked quickly.[129] This may be explained by the *pilum*'s short range. Modern re-enactors have found that reproduction *pila* have a range of between 10 and 20 metres. A Roman formation of perhaps eight men deep could not throw as a single volley because the depth of the formation is quite likely to be close to the range of the weapon. This would mean that the ranks would need to launch their weapons by successive ranks, requiring some preparation. Caesar describes how the Helvetii were disadvantaged by the *pilum* shafts encumbering their shields at the battle of Bibracte in 58 BCE and a volley of *pila* could pin enemy shields together.[130] Nevertheless, Caesar wrote that the Nervii returned some *pila*.[131] The *pilum* was a deadly weapon. A veteran soldier killed a general's horse with a single cast.[132] Roman soldiers continued to use the kind of sword described by Polybius which could cut with both edges and stab with the point, i.e., *gladius hispaniensis*, the Spanish sword. Most surviving Roman swords date from the Imperial period, although a number dating from the republican period have all the characteristics of the sword described by Polybius with a variety of minor design differences.[133]

Roman soldiers in Caesar's time used defensive equipment, such as shields, helmets and possibly armour, as they did in Polybius' time. Against his description of the Roman shield, a single example of a shield, which may be Roman, has been found in Egypt. The shield differs in construction details, but it is so similar that we may accept that Roman soldiers in Caesar's time used very similar shields to those described by Polybius.[134] Caesar's men apparently carried their shields in covers which inconvenienced them when the enemy attacked quickly, leaving no time to remove them.[135] Soldiers' helmets in Caesar's time were probably very similar to those Polybius saw. The monument known as the Altar of Domitius Ahenobarbus shows a helmet design referred to as Montefortino and a soldier, taken to be a cavalryman, with a Greek Boeotian-style helmet but no helmets found in archaeological sites of this period can be clearly identified as Roman.[136] The monument also shows Roman infantry much as Polybius' describes them (see Figure 12). While there may have been different crests, the helmets in Caesar's and Polybius' time are likely to have been very similar. At least some Roman soldiers marched between battles carrying rather than wearing their helmets, so Caesar had to order the soldiers to put their helmets on prior to a battle.[137] Some Roman soldiers still wore greaves, as Polybius describes. A relief from Urso (mod. Orsuna) in Spain shows two figures, both of which wear greaves (see Figure 14).[138]

As regards armour, a proposed restoration of the severely damaged Aemilius Paullus Monument in Delphi, dating from Polybius' time, suggests Roman infantry wore both muscle cuirasses and what is often called 'chainmail', more correctly 'ring mail', from its construction.[139] The figure in the muscle cuirass may well be an officer. Polybius tells us that wealthier infantry wore mail. Fragments of ring mail have been found in Roman sites, dating from the Republican period in Spain.[140] But how widespread was ring mail? Historically, ring mail has to be 'woven' from thousands of handmade rings, at least half of which must be riveted closed, if alternating rows of solid and riveted rings are used; otherwise, all rings must be riveted.[141] Modern experiments suggest that it would take 40,000 rings to produce a ring-mail coat. It has been estimated that it would require 200 days to produce the rings and 30 days to assemble the coat.[142] This armour was very expensive and probably handed down or sold when no longer needed.[143] While it may have been possible for large amounts of ring-mail coats to have been made over centuries, the large armies of the civil and social wars would have quickly used all of the supply. As Polybius' description of the armour of the *hastati* and *principes* shows, there may have been other kinds of armour used by the Romans in this period. A possible circular breastplate has been found during excavations at Numantia.[144] There are references to other kinds of armour. In the Urso relief (Figure 14) showing two figures referred to in the last paragraph, the right figure is not shown wearing mail; folds are shown, suggesting fabric. An incident at Dyrrachium during Caesar's campaign against Pompey suggests that even in Caesar's successful and comparatively rich, slave-owning army, many, even most, soldiers lacked armour.¹⁴⁵ When Caesar's troops occupied a position, rather than drive them off with a direct attack, the Pompeians used slingers and archers to harass them. This drove Caesar's troops to make up armour from leather, felt and textiles.¹⁴⁶ The grave stele from Padua of the centurion Minucius Lorarius of Legio III Martia in the Padova Musei Civici, dated to 44–42 BCE, shows him with his military belt, sword and staff but without armour (see Figure 15).¹⁴⁷ The well-equipped soldiers on some monuments may be more an ideal than a reality, an ideal readily embraced by modern re-enactors (Figure 16).

Polybius' comments on the importance of centurions as leaders within Roman armies are echoed somewhat in this period.¹⁴⁸ Caesar dealt with a morale problem by calling on centurions.¹⁴⁹ Later in Gaul, Caesar described the tactical control of a cohort by centurions and the bravado of two senior centurions.¹⁵⁰ Caesar believed that Domitius' army was effectively led by its centurions and military tribunes and the capture of a *primus pilus*, i.e. first centurion of a legion, was worthy of mention.¹⁵¹ Caesar notes the bravery of centurions in Spain, one of whom wore decorations in battle.¹⁵² Centurions led soldiers by example.¹⁵³ Cicero, referring to a reverse suffered at Amanus, specifically refers by name to a *primus pilus*, six other centurions and a military tribune who died when a cohort was destroyed.¹⁵⁴

Scholars continue to seek, as Delbrück did, to know how Roman armies were deployed and then fought.¹⁵⁵ The evidence is very limited as it is in earlier periods. Caesar used a triple line of veteran legions in Gaul but placed his newly raised legions and all his allied troops (*auxilia*) behind on higher ground.¹⁵⁶ In a battle between two Roman armies in Spain, we are told how they were deployed. Caesar's opponent, Afranius, had a double line of five legions with a third line described as allied troops, while Caesar's army of five legions had three lines with archers and slingers between the lines. Caesar's flanks were covered by cavalry.¹⁵⁷ This array is very similar to that described by Polybius except that Caesar's army is larger, the Italian allies are not separate and the heavier infantry are in cohorts rather than maniples. Roman troops appear to have still fought individually, as Polybius described. Caesar's Legio XII was so pressed together that the men were hampering each other in the use of their weapons. Caesar's response was to advance the standards and spread out the maniples within the cohorts.[158] Romans often charged at a run with a shout. Caesar criticized Pompey for suppressing this in his legions at Pharsalus.[159] Artillery is not treated in Polybius' description of Rome's armies but Caesar used artillery to support his battle line.[160] A cavalry commander in Africa was pinned to his horse by a bolt from a *scorpio* ('scorpion'), a type of catapult, panicking his men.[161]

Caesar's accounts of his armies' operations lay particular stress on the importance of the standards carried by each subunit. Polybius specifically describes the allocation of standard bearers, implying that they were important.[162] In his description of the landing on the British coast, Caesar notes that during the landing, his infantry has difficulty keeping their ranks and following their standards.[163] Caesar tells his readers that a sign of a disorganized, dispirited army was the gathering of the standards in one place and the soldiers not keeping their ranks and ignoring their standards.[164] Caesar used the standards to show his men where to form up.[165] In Africa, Caesar forbids soldiers moving more than 4 Roman feet, approximately 1.2 metres, in advance of the standards to reduce casualties from enemy skirmishers.[166] In this sense, the standards did not define the battle line of a unit which sounds more like a series of clumps than a line. The standards are mentioned in a number of versions of the *sacramentum*, the oath sworn by Roman soldiers. Polybius mentions the oath to obey commanders but does not provide a text.[167] Livy refers to an oath taken after the Battle of Cannae, claiming that this was the first time that soldiers were asked to swear such an oath. This oath included not leaving the ranks (*ordines*) other than to pick up a weapon, strike an enemy or save a citizen.[168] Dionysius of Halicarnassus has a similar version which refers specifically to the standards.[169] What this meant in practice is not entirely clear. It was evidently acceptable for tired and wounded men to retire within a unit, to be replaced by fresher men, as Caesar describes.[170] Even camp followers sought safety around the standards.[171]

There is evidence of local, contingent adaptation. During the conflict between Caesar and Pompey, Caesar's troops fought a particularly hard campaign against Pompeians in Spain in 49 BCE. Pompey's army had, it appears, adopted Spanish ways of fighting which included a much greater emphasis on skirmishing. Caesar explains that this occurs as a result of soldiers staying in a foreign land for a long time, an ancient version of the European colonialist 'going native'. Caesar reports that his troops, although they kept to their positions around their standards, were quite discomforted.[172] This is better understood when we learn later that Caesar's troops were under missile attack for five hours before charging, sword in hand.[173] Caesar even describes Pompeian troops as *caetrati* and *scutatae*, i.e. Spanish light infantry and close order infantry.[174] Caesar may well have simply been trying to blacken Pompey's name by claiming that his army was more Spanish than Roman but it is worth noting that, had Caesar's army had Polybian *velites*, it may not have been so vulnerable to skirmishers. Caesar's army in Africa had similar problems with skirmishers, requiring some retraining, i.e. adaptation to circumstances.[175]

Caesar's writings, particularly those actually written by him, i.e. *de Bello Gallico* and *de Bello Civili*, reveal an aspect of the Romans' armies not found in Polybius. Polybius, coming from the Greek military tradition, emphasized order, cohesion and control in the use of troops.[176] From what we can understand of Greek and Macedonian infantry equipment and organization, troops depended upon each other for mutual protection. As we have seen, Polybius claimed that a reason for the Romans' success against the Macedonian phalanx was that Roman soldiers fought as individuals. His description of Roman equipment reveals that the Romans placed a much greater stress on the use of missiles, whereas the Macedonians and Greeks restricted missile fire to lightly equipped troops that were not expected to fight hand to hand. Polybius' account of the Macedonian defeat at Cynoscephalae reflects his views of the relative advantages and weaknesses of Macedonian and Roman military arrangements.[177] Analysis of Caesar's language suggests a focus on the physical impact of his troops, seeking to raise it and to minimize factors, like terrain, which could lower it.[178] Caesar also uses the term *'animus'*, i.e. frame of mind or morale, constantly in his descriptions of his command of his army. This interest in the psychology of his soldiers is more prominent in his writing than that of the other ancient exponent of military psychology, the Greek general Xenophon.[179] This aspect of Rome's armies in this period is not found in Polybius, although his comment on what is required of centurions hints at what may have been a Roman perspective to which he gave little attention.[180] To some extent, Polybius' view is supported by Caesar's references to *virtus*, i.e. 'bravery' or 'manliness', often the resort of men who are in difficult positions.[181] Sallust's description of Metellus' victory against Jugurtha notes that the soldiers' *virtus* gave the Romans the victory, reflecting ancestral practice which accepted that generals do not always fight in the best circumstances.[182] Livy and Polybius both show a distaste for military trickery as opposed to reliance on *virtus*.[183] Nevertheless, from Caesar's works, it is clear that *virtus*, more in the sense of manliness than bravery, is a constant concern of generals and soldiers. So much so, a desire to demonstrate *virtus* could lead to disaster.[184] Roman soldiers in this period, while prone to substantial swings between despair, confidence and over-confidence, were strongly motivated by a desire to demonstrate their individual *virtus* as a cultural value.[185]

Caesar's armies present differently from those of Polybius principally because they are the armies of a state convulsed by civil strife, whereas Polybius' armies are those of a state that has survived threats to its very survival from which it emerged as the most powerful polity in the Mediterranean. We may speculate on the impact of that contrast on Rome's soldiers. The clear focus of a single authoritative source like Polybius is replaced by a multitude of different voices. In some ways our knowledge of Caesar's armies is more reliable because there are more sources against which we can sometimes compare accounts and different perspectives emerge, particularly from the Caesarian body of work. Battle descriptions depict large armies deployed in depth, using missile fire extensively with hand-to-hand combat. They did not manoeuvre greatly on the battlefield once deployed. Still, it is probable that soldiers and commanders from Polybius' armies would have been at ease in Caesar's armies across the two periods. In terms of equipment, if it is kept in mind that Polybius described a four-legion army, then the equipment of the soldiers may well be accurate. For the large armies of the Social War, Lucullus', Caesar's and Crassus' campaigns and the civil wars, much of that equipment may have been spread thin. A *legionarius* would have been more likely to have had a belt, sword, shield, helmet and *pilum*; additional armour would have been unlikely. Aristocratic officers, on the other hand, would have looked much more like the soldiers depicted on the near-contemporary monuments.

Chapter 8

The Armies of Imperator Augustus

Introduction

In this chapter, a further problem with the commonly accepted metanarrative of Rome's armies will be addressed. We will look at the impact which Augustus is claimed to have had on Rome's armies. The attention given to the organization and equipment of the armies will not be our focus; in most, if not all respects, Augustus' armies appear to have been identical in organization and equipment to the armies of Caesar. When we come to Augustus, a new problem must be faced. Augustus is commonly seen as the architect of the armed forces of Imperial Rome. There is remarkable unanimity among modern scholars regarding Augustus' effect on the armies and the changes, often called reforms, for which he is believed to be responsible.[1] Not all scholars use the reform discourse. Augustus is credited by Delbrück as giving the 'Roman Army' its 'definitive form' and an 'army system' (*Heeressystem*) by von Domaszewski.[2] In 1938 Cary referred to Augustus' 'military reforms'.[3] In 1959 Scullard listed Augustus' 'reforms' as:

- The creation of a 'standing professional force',
- The formation of a praetorian guard,
- Retaining twenty-eight legions as 'permanent units' with a fixed establishment,
- The creation of 'a second main branch of the army consisting of the Auxilia',
- 'The army was permanently stationed in the provinces',
- Defining 'conditions' of service including 'good pay', pensions and 'reasonable prospects of promotion',
- The navy was maintained 'as a regular force with naval bases'.[4]

In 1984, Keppie's description of Augustus' contribution to the Roman army accepts this view.[5] Roth in 2009 nuances the view slightly, sharing the changes between Augustus and the other Julio-Claudian emperors, but sees Augustus as mostly responsible. He also includes an additional change: that *legionarii* could not marry while members of the *auxilia* could.[6] In 2021, Whately sees that 'Augustus also carried out a vast number of reforms, some of which pertained to the military'.[7] This view of the influence of Augustus on Rome's armies is confined neither to generalist historians nor to the English-speaking world.[8] Augustus is credited with the 'creation' of the imperial army by Le Bohec's translator.[9] Mattern describes the armies at the end of Augustus' life compared to the previous century: 'For now we should note that by the end of Augustus' reign, the army had a different character'.[10] Junkekmann, while still accepting an 'army reform of Augustus' (*Heeresreform des Augustus*), sees it as an improvization.[11]

Augustus' armies and his political activities cannot be separated.[12] The views of Augustus as an army reformer are intrinsically connected to narratives of his principate. There is an ongoing debate in modern scholarship on the nature of Augustus' actions during his principate, including his military actions and even the very idea that Augustus could conceive of 'reform' is disputed.[13] Scholars have seen Augustus as carefully introducing a new form of government to Rome.[14] An implication of this view is that Augustus engaged in proactive planning to achieve his long-term aims. However, others have seen Augustus' success as the product of successful, reactive responses to crises or as evolutionary adaptations to circumstances.[15] It cannot be denied that Augustus was the most influential political figure in Rome from 30 BCE to his death in 14 CE. This domination had a great influence on Rome's armies because of the unbroken series of commands which gave Augustus control of almost all Roman forces. Before Actium, generals' practices affected mostly only their armies. When these generals' commands ended and their armies were disbanded or transferred to new commanders, the practices of former commanders no longer applied. Augustus commanded his armies within this framework but with a crucial difference. From

27 BCE, this single general directly controlled all Roman forces except for units in Africa, Illyricum and Macedonia.[16] Even more importantly, when Augustus died, his fellow pro-magistrate in a final joint command, Tiberius, retained the entire command intact and unchanged so that Imperator Augustus' command extended beyond his death to Imperator Tiberius. The outcome was that Augustus' armies became the only armies of the Roman people, so that any changes made by Augustus had a more universal effect compared to any previous Roman commander.

Augustus was not a 'reformer' but a Roman general with the largest command hitherto enjoyed by any Roman. He pragmatically addressed problems, the most significant of which was how to fund his armies. By the end of his life, his armies were the result of contingent adaptations to circumstances as he sought to achieve more than any Roman. To privilege these as 'policies' which determined the future shows a lack of appreciation of the degree to which Roman government, or indeed any government, is marked by evolving responses to particular problems rather than the establishment of future-orientated arrangements.[17] To deny that Augustus 'reformed' the military forces he commanded begs the question: what impact did Augustus have and how can it be explained? Augustus' military achievements can be better understood as a series of commands, not all of which were successful. The one significant change introduced by Augustus was to find a way of paying his soldiers through the *Aerarium Militare*. But even this change was by no means the result of prospective foresight: it was a pragmatic solution to an unintended, unforeseen consequence of his long, successive commands over the same troops.

Augustus and Change in Rome's Armies: Evidence of Reform

Augustus is credited with the decision as to how many legions would be maintained. Webster, for example, saw Augustus' most important decision to be the size and distribution of the army.[18] Until Augustus' time, the Romans lacked standing armies.[19] It cannot be doubted that Augustus posted legions and auxiliary forces to different provinces just as Pompey did when he had his Eastern command. The obvious difference is that Augustus' command did not end and, unlike Pompey, he did not need to disband his forces. Tiberius continued to command Augustus' armies following his death. As for 'standing armies', referring to forces in being between wars, the term is quite inappropriate for Augustus' armies, except perhaps the army in Syria. The armies in Spain, Egypt, Gaul and the Danube valley were still either expanding the area of Roman control, securing recently conquered areas or suppressing revolts. In Syria, the threat of a Parthian invasion and the need to be able to support Roman interests in Armenia meant that it was less an occupying force and more a constant restraint on Parthian activity.

It is claimed that Augustus created the Praetorian Guard, though it is acknowledged generals had previously had guards.[20] At this point, it is wise to be careful of our language. The body which we call the Praetorian Guard was known in Augustus' time as *cohortes praetorii*, praetorian cohorts. The 'guard' has echoes of more modern royal and imperial guards, e.g., Napoleon's Guard or the British Guards. The earliest Roman general, that we know of, to have a *cohort praetoria* was Scipio Africanus, according to Festus.[21] Following the assassination of Caesar, Octavian and Antonius both had praetorian cohorts.[22] After the Battle of Philippi in 42 BCE, 8,000 soldiers elected to remain enrolled and were assigned to praetorian cohorts for Antonius and Octavian.[23] Antony's praetorian cohorts were commemorated in a coin which he issued.[24] Orosius refers to five praetorian cohorts who accompanied him at Actium.[25] When Octavian returned to Rome, his praetorian cohorts accompanied him and remained with him for the rest of his life. Regardless of what can be said about the praetorians under subsequent emperors, Augustus did not establish praetorian cohorts but he retained them, although not all were based in Rome.[26]

In modern works, Roman armed forces of the Empire are particularly distinguished from that of the Republic by an organization in which the legions were joined by the *auxilia*. Mommsen believed that the introduction of the *auxilia* was a move away from the exclusive employment of Roman citizens.[27] Kromayer and Veith saw Augustus as making a major change in the *auxilia*. They saw this as the introduction of a fixed organization with standardized arms and equipment. This is presented as a measure to systematically exploit the militarily capable, non-Italian populations in the provinces. While it is acknowledged that the Romans used non-Roman troops prior to Augustus, he is credited with incorporating such units institutionally into Rome's armies. These units of *auxilia* are seen as permanent units of a 'standing army'.[28] Saddington also sees Augustus introducing 'regular or professional regiments in the *auxilia*', though, significantly, he does note the need to consider what professional means.[29] Webster also credited Augustus with reorganizing the *auxilia* and giving it 'regular' status. However, he believed that the regularization of the *auxilia* was 'introduced by stages throughout the first century'.[30] By 'regularization', modern authors refer to units being permanent, i.e. not temporary formations raised for a particular purpose. The term '*auxilia*' is, in fact although often unrecognized, a shorthand for one of the requirements of Rome's allies, to provide soldiers when required, i.e. *auxilia sociorum*, 'help of the allies'. Until the extension of Roman citizenship throughout Italy in the aftermath of the Social War, Rome's armies often had allied troops, mostly Italians.[31] After the Social War, allies outside Italy still provided troops, i.e. *auxilia*, prior to Augustus, although details are sketchy.[32] The attraction of non-Roman allied troops was that they both avoided the political problems of having to enlist Roman citizens and they were probably cheaper to pay.[33] The earliest units in Augustus' time had the name of their commander. By the reign of Vespasian, cohorts of these troops like I and II Flavia Brittonum existed but not in Augustus' time.[34] Names like these may reflect long years of service after which diplomas of discharge were issued from Claudius' time but nothing links Augustus to such naming conventions.[35] Cheesman found it difficult to explain why some units of *auxilia* were named after individuals rather than having the 'regularized' titles.[36] The simplest explanation is that the ethnic unit titles came into use long after Augustus' death, i.e. the use of tribal names for units of *auxilia* was an evolutionary change, without an author or reformer, without Augustus. Augustus did not create the *auxilia*. He certainly made use of allied troops, but the Romans had done so for centuries. After Augustus' death, units of the *auxilia* appear to have become permanent units in Rome's armies. This is yet another example of contingent adaptation that led to change.

It is commonly claimed that Augustus permanently deployed the legions to the provinces. Of course, prior to Augustus, Rome's armies were in the provinces or moving to or from between them, reflecting the commands to which magistrates with *imperium* had been assigned. In fact, the only other times when the armies had not been in the provinces were times of civil discord. As will be described later, from 27 BCE, Augustus has a series of commands, most of which he managed through his *legati*, i.e. representatives, often relatives. These commands, like all Roman commands, required soldiers. Revolts from Roman rule required armies to remain in provinces to suppress them. To claim a significance in Augustus' distribution of forces ignores the decisions of subsequent emperors. There is nothing to suggest that these men felt bound by Augustus' deployments. Their significance is only in relation to Augustus' military activities, rather than strategic insight or a decision on permanent deployment.

Discussion of 'conditions of service' reflect modern military arrangements. In preparation for my own commission in the Australian Army, I was required to study military law, enabled by the Defence and Army Acts which applied in 1973. These arrangements were inherited from the United Kingdom which has similar Acts which indeed serve to separate civilians from soldiers, and they are common in developed nations today. Military law was often implemented in the form of *Field Service Regulations* for British forces in the twentieth century. Delbrück believed that Augustus set the conditions of service through what he termed *consititutiones*: 'The Roman army received its definitive form through a comprehensive, systematic set of regulations proclaimed by Augustus, the *constitutiones Augusti*' to which Vegetius refers.[37] Phang believed that Augustus 'rationalized the conditions of service' but goes further, applying a 'Weberian concept of discipline', arguing that:

> 'Augustus and his successors and the Roman governing classes (who were also the literary classes) sought to routinize the professional army and to legitimate it through *disciplina militaris*.'

This included the deliberate separation of soldiers and civilians.[38] One further aspect of conditions of service is a ban on soldiers contracting legally binding marriages under Roman law.[39] Phang believes that Augustus prohibited soldiers from marrying although the only evidence of a ban is from Egyptian papyri after Augustus' death.[40] This ban is seen as a component of the reforms Augustus introduced for his 'new' army.[41] The problem with the claims for Augustus' conditions of service or *constitutiones Augusti* is that we have little or no idea of what they covered. We only know of their existence from the reference in Vegetius to Augustus', Trajan's and Hadrian's *constitutiones*. Vegetius' context refers to training of recruits. While there are extensive references in later legal codes referring to soldiers, there is nothing to suggest that Augustus' soldiers were under conditions very different from those described by Polybius almost 200 years earlier. In any event, as we saw earlier, Roman military discipline appears to be quite different from our expectations of soldiers today.[42]

Augustus has been credited with the maintenance of fleets in dedicated bases at Misenum and Ravenna. Star argued that the origin of the navy of the Empire is to be found in the Roman response to Mithridates in 88–84 BCE.[43] However, during the triumviral struggles, much larger naval forces under Sextus Pompeius attempted to force Octavian to negotiate from 43 BCE, culminating in the Treaty of Misenum in 39 BCE. This truce lasted until the naval battle of Mylae in 36 BCE which destroyed Sextus' power. The ultimate battle was fought at Actium in 31 BCE. For Augustus, naval power had been critical to his survival and success. Courtois believed that he maintained fleets for his own security in Italy, consistent with his experience in the civil wars.[44] Star argued that, while the evidence is lacking, Augustus had other fleets maintained in the Mediterranean, suggesting broader imperial motives.[45] Pitassi suggests that Augustus used his ships to limit piracy, understandable given his experience of the depredations of Sextus Pompeius' ships, but also for offensive operations. These included support for Augustus' armies operating in Germany and an expedition down the Red Sea.[46] There can be no doubt that Augustus had fleets and that he used them after 27 BCE when he received his first great command. In that sense, he is little different from Sulla, Pompey, Caesar or Antony, all of whom used fleets in military operations; Augustus cannot be seen to be an innovator in naval affairs. His repeated commands meant that, on his death, the fleets continued to serve under Tiberius as part of their joint command.

The 'reforms', as they are often termed, claimed for Augustus mostly reflect assumptions arising from what is known of Roman armies after Augustus' death. To claim that Augustus intended to 'reform' the armies lacks an evidentiary basis. The application of the discourse of 'professionalism' for Rome's armies prior to Augustus is unsustainable.[47] Augustus' soldiers certainly served for long periods but there is nothing to suggest that they were enlisted under different conditions or with expectations different to those under which Roman soldiers had served for hundreds of years. Long service might produce efficiency but is does not necessarily amount to the modern construct of professionalism.[48] Delbrück's *constitutiones Augusti*, if they existed, have not survived. Evidence of a ban on marriage post-dates Augustus. Augustus used fleets as did earlier Roman commanders. Augustus had an impact on Rome's armies but it was indirect, unplanned and unintended. Among these unintended changes were the 'undying armies' which extended military service for Augustus' and Tiberius' soldiers, as we have already seen in Chapter 2.

Augustus' Achievements and Failures

While it is possible to dispense with the 'army reforms' ascribed to Augustus, it begs the question of what his achievements were in relation to Rome's armies. Suetonius states of Augustus that 'in military things, he changed and put in place many things' (*in re military et commutavit multa et instituit*). He continues: 'and also brought back several things to old customs' (*atque etiam ad antiquum morem nonnulla revocavit*). The first item in support of these claims offered by Suetonius is *disciplina*, i.e. discipline, not a 'reform' but a common trope. Restoration of earlier practices can hardly be styled as reform.[49] It should be noted that Suetonius does not claim that Augustus put anything new in place.

Augustus himself left to posterity his view of his achievements. His *Res Gestae Divi Augusti* ('The Achievements of the Divine Augustus', *RGDA* for short)) includes claims of two kinds in relation to the armies: conquests and rewards for soldiers. Augustus claimed that he increased the limits (*fines*) of Roman territory where people lived who were not subject to Roman authority (*imperium*).[50] He further claimed that he pacified (*pacavi*) the Gallic and Spanish provinces and Germany from Gades in Spain to the mouth of the Elbe.[51] He claimed that he added Egypt to the control of the Roman people and recovered standards (*signa militaria*) lost by others in Spain, Gaul, Dalmatia and to the Parthians.[52] He made a specific claim of conquering the Pannonians, extending the limits of Illyricum to the banks of the Danube and crossing into Dacia to win a victory and force the Dacians to accept Roman commands.[53] Regarding soldiers, Augustus advertised the number who swore the soldiers' *sacramentum*, the oath to obey him, and the number whom he placed in colonies or allowed to return home.[54] Augustus made a very specific claim about how he found the land for the soldiers he released from military service: he paid *pecuniam pro agris*, i.e. money for the land, listing seven separate occasions.[55] He claimed to have provided the advice to establish the *Aerarium Militare*, i.e. 'Military Treasury', and to have contributed money to it.[56] He also claimed to have founded colonies for soldiers in Italy and in provinces.[57] Separately, Augustus claims to have given 2,400,000,000 sesterces to the treasury or the Roman plebs or to discharged soldiers.[58] He uses the word *dismissis*, the same word Livy used of Ligustinus' discharges. Of these claims, the detailed payments cannot be tested other than by the silence of contemporaries who did not question these very public statements; the *Aerarium Militare* is known. As for the conquests, these can the traced though the surviving sources. However, it ought to be noted that Augustus made no claims for any change in relation to the armies other than for the establishment of the treasury to pay the soldiers. Unlike his adoptive father, he rarely commanded armies personally, but he certainly engaged closely with their activities.[59]

Spain, Recovering the Lost Standards and Release of Troops

Following the deaths of Antony and Cleopatra, Augustus' first task was to disband the armies which fought at Actium. In 30 BCE Octavian found himself in control of a vast number of men who had to be paid. This army amounted to perhaps sixty legions and it should not be forgotten that many were from Cleopatra's and Antony's forces. It also included troops provided by *socii*, i.e. *auxilia sociorum*, 'Auxilia'. The soldiers provided by Eastern client rulers would certainly have been seen by both Octavian and Antony as *auxilia sociorum*, though the status of Deiotarus' forces, equipped and trained as Roman *legionarii*, is uncertain.[60] There was also a substantial fleet and its crews. Just as with every previous war, this vast force had to be, to use a very modern term, demobilized, because without a new war and new booty it could not be paid, and it was potentially politically dangerous.[61] Augustus may have decided not to retain married men. Many may have been pressed and they would have been less attracted to remaining after the end of the civil war brought an end to rich booty. The acquisition of Egypt by Augustus provided funds to pay them out.[62] Augustus describes this in the *RGDA*, referring to the settlement of 300,000 men in colonies or returned to their homes.[63] Yet much of the army was retained for reasons which require explanation. Why were so many legionary and allied troops retained? There are two reasons which are consistent with the evidence and the circumstances: internal challenges to Augustus' domination of affairs and the requirements of foreign wars.

It may have taken three years to manage such a massive demobilization but by 28 BCE the normal processes of Roman government appear to have been operating. Augustus took the credit for this, surrendering an informal, de facto control to the senate and people of Rome.[64] The circumstances are obscure but Augustus was given a large *provincia*, i.e. province or area of responsibility, which covered most but not all the areas where triumphs could be won. This new command covered everywhere under Roman control to which armies were allocated except Africa, Illyricum and Macedonia. There is no reason to believe that those provinces not under Augustus command had troops and commanders assigned to them in any way different from how they had been for centuries. This explains why Crassus was able to fight a war, separate from Augustus, in his Macedonian province against the Bastarnae and make a claim for the *spolia opima*, the 'supreme spoils' awarded for having personally defeated the enemy chief in battle.[65] It is assumed that, under Augustus' influence, Crassus had his claim for the *spolia opima* rejected, but he could not be denied a triumph.[66] What is interesting is the effort to which Augustus went to ensure that Crassus did not have his claim for the *spolia opima* met. Augustus claimed to have found an inscription on a 400-year-old linen corslet that said that Cossus was a consul when he won the *spolia opima*; because Crassus was not consul when he defeated Deldo, Augustus opposed his claim. Livy was far from comfortable with Augustus' views on this matter.[67] Livy was aware that in literary sources Cossus had been described as military tribune. And his audience might well have wondered at the difficulties of reading a 400-year-old linen garment.[68] Another general's claims to victories, this time from an area controlled by Augustus, also caused problems for Augustus; Cornelius Gallus' activities in Egypt compelled Augustus to renounce his *amicitia*, i.e. friendship, what we might call a political kiss of death.[69]

Augustus, as consul from 27 to 23 BCE had command over most of the Roman provinces which had armies and scope for conflict.[70] This provided Augustus with an opportunity to win laurels by defeating non-Romans and expanding the *pomerium*, the symbolic sacred boundary of Rome, as his adoptive father Julius Caesar had done. In 23 BCE, when Augustus was very ill and laid down his consulship in obscure circumstances, he was given a five-year command to complete the work he had begun. This *provincia*, with some changes in its territorial constituents, remained his for the rest of his life but the laws which gave it to him had to be passed for each renewal as was required by Roman precedent. Augustus, through his *imperium* as the appointed magistrate, was able to appoint *legati*, as Caesar and Pompey had done before him, and he used these *legati* where he did not attend to matters in person. From this time until the destruction of Varus' forces in Germany, Augustus embarked on the largest programme of territorial expansion the Romans had yet seen. In comparison to the Gallic campaigns of his adoptive father, Augustus deployed more troops in more areas to eventually secure territory stretching from Spain to Pannonia. He also managed to have the Parthians return the standards lost at Carrhae and the others lost by Antony's generals. All these campaigns were to have a substantial effect on the armies by extending the length of the soldiers' service.

In 26 BCE Augustus began a campaign to expand the Roman province of Hispania Tarraconensis, the second of only two campaigns in which he commanded troops directly, though only for a short time.[71] This campaign lasted under other commanders until 19 BCE and required six legions, three of which remained in Spain, presumably to secure the conquests.[72] If Velleius' view of the significance of the campaign is any indication, the victory was noteworthy to contemporaries.[73] Its aim was to suppress the Spanish hill tribes which appear to have periodically raided the areas under direct Roman control. Augustus initially deployed three legions: Legio VI Victrix, Legio X Gemina and Legio V Alaudae.[74] Later three more legions were added: I Augusta, II Augusta and IV Macedonica, though it is not clear when. The name of another legion, IX Hispana suggests that it was in Spain before its deployment in Illyricum.[75]

The campaign in Spain was a significant undertaking and it lasted for years with several revolts.[76] Augustus eventually used almost as many legions as Caesar had in Gaul. The campaign was complicated because two different tribes were engaged at the same time.[77] While he was in Spain, Augustus had Terentius Varro suppress the revolt of the Salassi in the Alps in 25 BCE. Augustus marked the successful end of these two wars in two ways. As with previous Roman commanders, he discharged troops. At the conclusion of the hard-fought campaign against both Spanish peoples, Augustus discharged aged troops to build what was probably a military colony in Lusitania, Augusta Emerita.[78] When the Salassi in the Alps were defeated, the tribe's best land was used to house discharged praetorians in a new colony, Augusta Praetoria.[79] Augustus also closed the doors of the Temple of Janus in Rome to show the peace his command had brought the Roman world.[80] This gesture showed the wisdom of those people who gave him his extensive command and was an achievement in itself, but the closing of the temple in 25 BCE was premature. The Cantabri and Astures in Spain revolted, only to be suppressed and to revolt again and be suppressed yet again, showing the reason that troops had been left there. In Egypt in 26 BCE, Aelius Gallus mounted an expedition into Arabia and the next governor, C. Petronius, repelled raids by the Ethiopians.[81] Regarding the troops, Augustus' discharge of the praetorian veterans and other troops was no different from what other Roman generals had done at the end of a successful campaign.

Following in the footsteps of many Roman commanders in the past, Augustus went to the East in 21 BCE. The Parthian king, Phraates, was persuaded to return the lost standards of Crassus and Antony to Augustus.[82] This was the result of diplomacy but one cannot discount the effect of the implied threat of the Roman legions in Syria on the Parthians. The potential for a Roman invasion and the restoration of Tiridates, Phraates' son, were sufficient encouragement.[83] Augustus also arranged matters in the East, maintaining subject rulers in their states and directing the young Tiberius to secure Armenia for Tigranes, a pro-Roman prince.[84] While Augustus remained in the East, Agrippa was sent to Gaul to settle a matter of German raiding but, having dealt with the Germans, he was called to a further Cantabrian revolt in Spain in 19 BCE. While he suppressed this with difficulty, he had a different challenge with soldiers who demanded discharge on the basis of age.[85] Evidently Agrippa persuaded the troops to continue in their service with a combination of threats and encouragement. This was the first of a number of problems Augustus had with troops demanding discharge and suggests that his soldiers believed that their terms of engagement were no different from those of Roman soldiers in past centuries. These men had fought many campaigns and survived but the stubbornness of the Cantabrians drove them to seek discharge rather than continue to fight a stubborn but probably impoverished enemy.[86] On this occasion, we do not know how Augustus eventually addressed the concerns of the soldiers, but we may suspect that the soldiers felt that his generosity in settling men in Augusta Emerita was not a sufficient reward. There is a reference to Augustus discharging troops without rewards when they demanded discharge. This may have occurred at this time.[87]

Germany, the Danube Valley and Paying the Troops

At the end of Augustus' first ten-year command in 18 BCE, Dio indicates that Augustus received a further five-year extension of his command in his *provincia* and Agrippa received similar powers as a co-magistrate. Augustus had suggested that five years would be enough to settle affairs.[88] It is not hard to imagine that there was concern within the aristocracy that he was monopolizing military commands, but Augustus wished to achieve greater honours. He turned his attention to Gaul, and it is possible that challenges to Roman authority in this area were the basis of his request for an extension. In any event there was a revolt among Alpine tribes. This was suppressed but was followed by disturbances in Pannonia and Dalmatia, which were to a degree provoked by Silius Nerva's campaign in the Alps in 17/16 BCE to suppress the revolt there. There were also raids in Macedonia and trouble in Thrace. There was a substantial raid into Gaul in 17 BCE by three German tribes, the Sugambri, the Usipetes, and the Tencteri. However, the Germans made peace before serious campaigning began.[89] This German raid was serious enough to result in the loss of the eagle belonging to Legio V under Lollius, an event described by Suetonius as 'more notorious than damaging' (*marjoris inflama quam detrementi*).[90] Rhaetian raiders were also causing problems in Gaul.[91] Augustus' solution was to send his *legati*, Tiberius and Drusus, to attack the Rhaetians from the Alps and

Gaul, a mission they completed in 15 BCE.⁹² In 14 BCE there was a further revolt in Pannonia and a revolt by the Comati in the Alps.⁹³ Augustus' success in the Alps is commemorated in the monument at La Turbie of 7/6 BCE.⁹⁴ It lists forty-six Alpine and Rhaetian tribes conquered by the Romans under Augustus' auspices.

Dio records that in 13 BCE, Augustus returned to Rome. A hoarse Augustus had a quaestor read out a list of his achievements and a proposal to the Senate that defined the number of years citizens should serve in the army and how much money they should receive on discharge as an alternative to the land they were always demanding.⁹⁵ The reference to how long troops should serve is intriguing. Citizens who had remained in the legions in 31 BCE would now be approaching their twentieth year of service. Polybius records the expectation of service as up to twenty years.⁹⁶ While some men had been released into veteran colonies, perhaps Augustus was concerned that he might not be able to pay the benefits his soldiers may have been expecting. The wars Augustus had fought were not in areas which might have provided rich booty. The presentation to the Senate is significant because it coincides with the end of the five-year extension of his command. What Augustus had read to the Senate sounds very much like a rendering of an account of achievements combined with the presentation of a problem and a solution: the regulation of the discharge from service of Augustus' soldiers. We know from *RGDA* that some troops were released from service in 14 BCE and settled on lands in Italy and the provinces.⁹⁷ Dio's judgement, perhaps reflecting his own experience many years later, is that the population were relieved that they would not be robbed of their possessions to pay the troops.⁹⁸ Augustus is insistent in *RGDA* that he 'paid money for land' (*pecuniam pro agrisi*).⁹⁹. The settlement of troops on land would be entirely consistent with the end of a command as it had been for Pompey and other generals in the recent Roman past. The discussion in the Senate, however, of how the discharges of his soldiers from his army were to be funded, and Dio's observation, indicate what might be the first phase in the establishment of a means of paying discharges of soldiers in armies composed of soldiers from 'undying' armies. Whereas the question of the payment of soldiers in the past would have been implicit in a general's command, the vast numbers of Augustus' soldiers made provision for their rewards on discharge a broader problem. Although Augustus' command was extended to 8 BCE, contemporaries, based on precedent, would have expected that all the soldiers would be released from service when Augustus' commands finally came to an end.

Following the extension of Augustus' command for a further five years to 8 BCE,[100] Agrippa was sent by Augustus as *legatus* to Pannonia.[101] This may indicate the need for Augustus' extension: instability in the area south of the Danube and a threat to Dalmatia but it may have signalled the intention to secure all the territory south of the Danube between Rhaetia and the Black Sea. The conquest of the territory between the areas under Roman control and the Danube was as substantial an undertaking as Caesar's conquest of Gaul. What is surprising is that Augustus should have attempted it at all. The only link between the new conquests and the business of the previous two commands was the attack on neighbouring Rhaetia. While it is possible that that operation might have caused some instability to the east, it still does little to explain the decision. It is most likely that Pannonia was chosen simply because of its unconquered nature and the *gloria* to be had by winning it and just as Caesar had advanced to the Rhine, Augustus had advanced to the Danube. In his *Res Gestae*, Augustus makes much of the achievement.[102] Also, the opportunity for the winning of booty with which to reward the troops should not be discounted. The decision to undertake this campaign shaped the military activity of the remainder of Augustus' life and provided a first practical reason why Augustus' soldiers were still undying: they were still obligated to serve the magistrate in whose command they served.

The invasion of Pannonia in 13 BCE was not the only new undertaking. While the Pannonian campaign was being fought, Drusus, Tiberius' brother, began in 12 BCE what was to be an entirely new series of campaigns in Germany against the Sugambri and their allies. The campaign may have aimed at nothing less than the advancement of the area under Roman control from the Rhine to the Elbe.[103] Some have assumed that Augustus saw the Roman frontiers in modern strategic terms. Scullard epitomizes this approach where he identifies a specific policy for Augustus on the northern frontier at this time:

> In order to secure Cisalpine Gaul Augustus decided that it was necessary to conquer the whole Alpine range and by reducing the districts of Raetia and Noricum (roughly eastern Switzerland, the Tyrol and Austria) to advance the frontier to the Danube from Lake Constance to Vienna. He further judged that the southern Balkans would never be secure unless the Romans advanced northwards there also to the Danube.[104]

There are a number of questionable assumptions behind the belief in such an approach by Augustus.[105]

Drusus again crossed the Rhine and established camps north of the Rhine. Such camps indicated an intention to continue campaigns in Germany.[106] A series of campaigns followed which permitted Drusus to reach the Elbe in 9 BCE.[107] Even after Drusus' death, the campaigning in Germany did not cease.[108] Tiberius was sent as a replacement until 7 BCE and others continued the work. At the same time, there was a revolt in Dalmatia. The impact of the invasion of Pannonia was to make Dalmatia more rebellious and it was added to Augustus' *provincia*.[109] An internal revolt in the client kingdom of Thrace in 11 BCE drew Calpurnius Piso, governor of Macedonia, to successfully suppress the rebellion.[110] Augustus attempted to close the Temple of Janus again but was unable to do so. This attempt to close the door of Janus' temple is dated by Dio to after Augustus' census in 8 BCE.[111] This would be at about the time that Augustus' five-year command would have ended. It would seem that Augustus was attempting to repeat what he had tried to do at the end of his first command when the temple was briefly closed. As it was, raids by Dacians across the Danube and a revolt by the Dalmatians caused Tiberius to move from Gaul.[112] Drusus had recently died and Germany may have been quiet. In response, Augustus was given another ten-year command in 4 CE. A justification for the command may be indicated by Tiberius's dispatch across the Rhine to continue Drusus' campaign.[113] Tiberius took the field in Germany again in the following year too.[114] Operations in Germany seem to have continued, though the details are unclear.[115]

In 5 CE the soldiers of Augustus protested about the length of service that was required of them and the rewards they received. For the second time, Augustus looked ahead to how he would be able to pay his ageing army. The result was the establishment of the *Aerarium Militare* in 6 CE.[116] The two key texts, which describe Augustus' arrangements for the payment of his troops, are those of Suetonius and Cassius Dio. Suetonius' text provides a very general description which stresses Augustus' aim to discharge the troops 'so that either due to age or poverty, after service, they are not tempted to revolution' (*ne aut aetate aut inopia post missionem sollicitari ad res novas possent*).[117] Dio on the other hand gives a fuller account which stresses the protracted nature of Augustus' approach to persuading the Senate that a fixed tax would be needed because he could not pay all his soldiers.[118] The initial funds came from contributions by Augustus and Tiberius, but it took some persuasion to get the senators to agree to a five per cent tax on inheritances and bequests to supply funds in the future.

But why undertake this measure now? We can only speculate. Given that troops were discharged after his conquest of Spain in about 23 BCE, this would mean that men enlisted to replace these would be at or past their twenty-fifth year of service in 5 CE. Men would have been engaged to fight the wars in Germany and Pannonia, and many would have been approaching their twentieth year of service. Such lengths of service had probably not been seen since the Second Punic War. Further, the wars in Germany and in Pannonia probably required more troops to be raised to fill out the legions; we should not assume that the legions were kept at maximum strength unless there was a need.[119] For Augustus, the establishment of the *Aerarium Militare* was the solution to an immediate as well as a long-term problem of paying his army as well as funding retirement benefits. His armies were not winning enough booty to be self-funding. Dio believed that Augustus had a solution but that it took some time to persuade the senators that something needed to be done. It does not take either accounting expertise or actuarial skills to see that if Augustus' campaigns continued and his soldiers were not disbanded, eventually a great many would have to be paid discharge bounties while it was necessary still to pay *stipendia*, i.e. salaries, to those still serving. There is evidence that Augustus had discharged numbers of troops in 7, 6, 4, 3, and 2 BCE but it seems that by 6 CE he had exhausted his resources.[120] The establishment of the *Aerarium Militare* was an indication of how much had been changed by the existence of Augustus' undying soldiers. It was an attempt, as Suetonius saw, in isolating this aspect of the decision, to address the problem of the need to pay discharges to soldiers in the future. The solution was a typically Roman contingent decision forced on an unwilling Senate by an ageing *imperator* with pressing problems. As Watson has observed, this decision 'was, of course, not the final stage of a far-reaching plan, but simply the solution forced on the emperor by the passage of events'.[121]

In the field Tiberius campaigned in Germany until the revolt of the Pannonians in 6 CE.[122] Tiberius was evidently attempting to attack the Marcomanni from the east while a force from Illyricum was attacking them from the south. It is clear that by this time the Romans had a grasp of the geo-political features of the area. Years of campaigning had been a good teacher. We can see this as a pincer movement, a metaphor which depends on a graphical representation of space, i.e. a map, but, because the Romans lacked the basis for the metaphor, i.e. maps, we should not expect the Romans to have seen it in the same way.[123] Significantly, the attack was on a tribe, an ethnographic entity, not a geographical feature. In any event, the Pannonian revolt, which broke out at this time, was so widespread that it disturbed all Roman arrangements south of the Danube. Tiberius was forced to break off his operations in Germany and move his forces to Pannonia to suppress the revolt.[124] The threat was so serious that Augustus believed that Italy might be threatened and a levy was instituted.[125] The procedure was probably similar to that described by Polybius. However, the immediate threat passed and the Pannonians were defeated by the Romans' Thracian allies and a stubborn Roman resistance. In further disturbances, Moesia was attacked by the Dacians and Sarmatians from across the Danube and Macedonia was raided, further complicating the task facing Tiberius. The Pannonians avoided pitched battles with the Romans and resorted to what we would see as guerrilla tactics.[126] Dio believed that Augustus sent an army with Germanicus because of distrust of Tiberius. This army included freshly levied freedmen, a measure of the threat posed by the Pannonian revolt.[127] By 8 CE the revolt was effectively over and the Pannonians sought peace, though some resistance remained.[128] Augustus had been sufficiently concerned by the revolt to move to Ariminum, a road junction in northern Italy, to supervise his *legati*.[129] After prolonged fighting, Germanicus and Tiberius restored Roman control by 9 CE.[130] It was as this time, 9 CE, that Varus' forces were destroyed in Germany in the *clades variana* (Varian disaster), with the loss of three legions and their associated auxiliary troops (Figure 13).[131] The site of the disaster has been found at Kalkriese in the Teutoburger forest.[132] This was the most serious reverse that troops under Augustus' auspices ever suffered.[133] Augustus feared that Italy would be open to attack.[134] His response was to institute a further levy, recall veterans and enlist freedmen.[135] He also removed Gauls and Germans to places where they could be monitored.[136] Tiberius moved to guard the Rhine to prevent any major incursion.[137] In the following year, 11 CE, Germanicus and Tiberius both crossed the Rhine but, significantly, avoided an advance into the interior.[138] This was a demonstration and it was effective, but only underlined the failure of Augustus' German campaign. It is difficult to believe Augustus' claim that he had pacified (*pacavi*) the Germans.[139]

In 13 CE, Augustus received his fifth command jointly with his adoptive son, Tiberius.[140] Given his advanced age and infirmity, it was really a command for Tiberius, and provided further opportunities for Germanicus and a means of advancement for the younger Drusus.[141] The cost of paying the soldiers on discharge was proving a burden to those subject to the tax which funded the *Aerarium Militare*, and a threat to the state. Augustus needed to look for other sources of funds. Dio indicates that Augustus' motive was not to abandon it but to persuade the wealthy that if no alternative could be found that seemed better, they needed to accept the tax as a permanent measure.[142] When an alternative was strongly proposed, Augustus accepted one and began to implement it in the belief, Dio tells us, that the five per cent tax would eventually be seen as less burdensome; and it was.[143] Augustus saw what he was doing as providing for greater security for the propertied classes, who would suffer if there were discord from the veterans. This was the third attempt to make provision for the discharges of the soldiers in armies of undying soldiers. Augustus manoeuvred the senators to accept the existing solution, another contingent arrangement. The *Aerarium Militare* was the product of a series of pragmatic decisions.

Death for Augustus, Birth of the Imperial Armies

Imperator Augustus, as Octavian became in 27 BCE, was the most successful *princeps* in Roman history by the time of his death in 14 CE. His measures of success, proclaimed publicly in *RGDA*, were his conquests and his ability after his victory at Actium to settle soldiers on land, thanks to the wealth of Egypt, without confiscations or proscriptions. While the armies which fought under his auspices were successful initially in Spain and in the East, the campaigns in Pannonia were expensive though eventually successful. The campaigns in Germany were a failure which resulted in the loss of three legions and an unknown number of allied soldiers. It should not be forgotten that campaigns were still being waged at the end of Augustus' life. This may be the context of Augustus' advice to Tiberius concerning the expansion of the provinces.[144] Augustus believed that, at the time of his death, it was inopportune to expand territory which, like Spain and Pannonia, would need to be occupied. Because Tiberius and Augustus jointly shared a command, all the troops, Roman and allied and the naval forces continued to serve. That this led to Dio's 'undying armies', effectively permanent forces, was an accident of history, not military reforms. Apart from the military successes and failures of Augustus, his long series of commands produced a problem hitherto unseen: the need to reward soldiers after very long service. Augustus appears only to have realized that there was a problem in 13 BCE. Prior to that, he had released soldiers at the end of a successful campaign in Spain. But the wars that followed were such that he could not release men. Roman soldiers had always served for booty, and the previous century of the acceptance of the *capite censi* into the legions had led to expectations that soldiers would be rewarded on discharge; the lack of booty from years of unprofitable campaigns would not have helped. The result was the creation of the *Aerarium Militare* and the grudging acceptance by the wealthy that they would need to pay for it; even Augustus' great personal wealth was insufficient. As we have seen, Augustus cannot be called a military reformer, unless in Suetonius' terms of restoring discipline and other traditional measures. His single innovation was a grudging one and more in the nature of a contingent adaptation to unforeseen circumstances. And even here, he retained at least some soldiers in their units well beyond their traditional twenty years. However, this should not lessen an appreciation of the achievements which he claimed for himself in *RGDA*. After all, apart from the creation of the *Aerarium Militare*, he never claimed the military reforms for which his is so commonly credited. Augustus died after a short illness in 14 CE. He provided a set of accounts to the senate, as would any Roman magistrate at the conclusion of his command although in Augustus' case, the command continued with his colleague, Tiberius.[145] Augustus' written direction to Tiberius, according to Dio, was not to expand the empire because it would be hard to guard. Whatever this meant, later emperors clearly did not feel bound by it, not even Tiberius whose armies continued to operate across the Rhine.[146] According to Dio, Tiberius did not assume that he would be emperor, though he accepted the inheritance of Augustus' property; they were not yet the same thing. He did not assume the title of *imperator* immediately.[147] Tacitus on the other hand presents Tiberius as taking control of affairs immediately.[148]

The reaction of the troops in Pannonia to the news of Augustus' death was to demand discharge, a new leader, new status and a new state.[149] Velleius, a contemporary describing this revolt, believed that the most outrageous action of the German and Illyrian legions was their attempt to tell the emperor how much they should be paid and the length of their military service.¹⁵⁰ For the troops, Augustus' arrangements had been less than successful. Tacitus describes the concerns of the troops in Pannonia.¹⁵¹ The soldiers' grievances may be summarized as follows:

- Their length of service was unlimited. Some men had served thirty or forty campaigns. Even discharged men were kept with units (*sub vexillis*).
- Land grants were in poor, swampy country.
- Military life was unprofitable, i.e. there was a lack of booty, with too many deductions from pay.

The list is instructive and consistent with Velleius' description of the demands of the legions. Evidently, the land grants of which Augustus boasts in *RDGA* were less attractive to the soldiers than to the wealthy whose property Augustus had been protecting. It would appear that Augustus was unwilling to release soldiers after very long service in spite of the strong link between the emperor and his soldiers.¹⁵² Perhaps the *Aerarium Militare* lacked the funds to pay the bounties though it could have been the pressure of the Pannonian troubles and, indirectly, of the *clades variana* on army manpower.¹⁵³ However, it is evident that the policies modern and ancient writers see as Augustus' creation were not fully implemented even by their author. The soldiers' demands were for sixteen years' service, like the Praetorians, and no service *sub vexillis* as a prolongation of their service.¹⁵⁴ The problem of long service was not confined to the Pannonian armies. In Germany, Germanicus subsequently found men with toothless gums and limbs bent as a result of old age.¹⁵⁵ There were men demanding release in the German army who had seen thirty *stipendia* or more, i.e. thirty or more years' service (*tricena aut supra stipendia numerantes*).¹⁵⁶ The troops were concerned about hard toil, poor pay but most of all, the veterans wished to be released from their long service, well beyond the term that Augustus claimed to have set.¹⁵⁷ The troops were discharged and their bounties were paid.¹⁵⁸ Later Tiberius had many men due for discharge because they were time expired but they were not released.¹⁵⁹

Conclusion

The aim of this book has been to challenge the commonly accepted views of Rome's armies from earliest times to the death of Augustus in 14 CE. This entailed tracing a quite modern history of the history of Rome's armies. In the 2,000 years since the death of Augustus, the manner in which we examine the military arrangements with which he was familiar is barely 150 years old. But this led to two additional tasks. The first has been to review metanarratives, the commonly accepted contexts, which are applied to the Romans' military arrangements in this period. Since the middle of the nineteenth century, these arrangements have been termed 'the Roman army'. The second metanarrative is the use of the term 'reform' to explain, even to describe, changes in the Romans' military arrangements detected in the surviving ancient sources. The third metanarrative is the use of 'professionalism' to describe a perceived social-political change in the composition of the armies. In the case of all three metanarratives, I have argued that their use is not sustainable. Not only do they lack an evidentiary basis, but they also carry implications which are, at best, anachronistic.

The second task has been to propose a different characterization of the armies by explaining changes as contingent adaptation by Roman commanders to the circumstances in which they operated, mostly producing no permanent change. While the earliest armies are effectively almost impossible to accurately describe before the outbreak of the First Punic War in 264 BCE, the subsequent armies are really defined by Polybius' rich description. The later information from Caesar and his near contemporaries does not, in fact, significantly change the picture. Even regarding equipment, there are few changes: helmets, swords, missile weapons, shields show little change. Even cohorts are described by Polybius. It could reasonably be claimed that had a military tribune from the Second Punic War been teleported into Caesar's army in Gaul he would have found that army to be quite familiar.

It is inevitable and reasonable for modern readers to look for what is familiar to them in these Roman armies. Modern armies are financially burdensome, technologically daunting and significant national institutions: we look to the Romans for the challenges that such institutions offer us. A particular deficiency for modern readers in ancient accounts of warfare is the almost total silence on logistics, particularly food supply, although its importance was recognized.[1] Rarely are such concerns mentioned in the surviving sources.[2] We know almost nothing about how and where weapons were made. For example, by what means were the sixty legions which Octavian demobilized after the Battle of Actium equipped? We are left only with educated guesswork. Similarly, we struggle to even imagine how the Romans actually fought their battles, lacking as we do the bird's-eye description that Thucydides gave us in his account of the Battle of Mantinea in 418 BCE, fought between two hoplite armies. Descriptions of Roman battles from Caesar, Polybius and Livy were meant for people who knew so much that we may only imagine. For historians, amateur and professional, the fascination may remain but, in our unending conversation between the present and the distant past, there must always be reflective caution distinguishing what is known, from what is assumed.

Abbreviations

It has become customary in Ancient History and Classical Studies to use standardized abbreviations for the works of classical authors and their texts and for some reference works. The abbreviations in this work as based on those in the *Oxford Classical Dictionary*. For the convenience of readers, the abbreviations of reference works and Greek and Latin works are listed below. *Ibid.* is an abbreviation for *ibidem*, i.e., in the same place.

Common Abbreviations for reference works

AE	*L'Année Épigraphique*
Cod. Just.	*Codex Iustianus*
CIL	*Corpus Inscriptionum Latinarum*

Abbreviation for author	Author	Abbreviation for work	Name of work
Amm. Marc.	Ammianus Marcellinus		
App.	Appian	*Hisp.*	*Spanish Wars*
App.	Appian	*Pun.*	*Libyan Wars*
App.	Appian	*B. Civ.*	*Civil Wars*
App.	Appian	*Mith.*	*Mithridatic War*
Arr.	Arrian	*Tact.*	*Tactica*
Ascl.	Asclepiodotus	*Tact.*	*Tactics*
Augustus	Augustus	*RGDA*	*Res Gestae Divi Augusti*
Aul. Gel.	Aulus Gellius	*NA*	*Noctes Atticae*
Caes.	Caesar	*BCiv.*	*De Bello Civili*
Caes.	Caesar	*BHisp.*	*De Bello Hispaniensi*
Caes.	Caesar	*B. Gall.*	*De Bello Gallico*
Caes.	Caesar	*B.Afr.*	*De Bello Africo*

Abbreviation for author	Author	Abbreviation for work	Name of work
Caes.	Caesar	B.Alex	De Bello Alexandrino
Cass. Dio	Cassius Dio		
Cic.	Cicero	De Or.	De Oratore
Cic.	Cicero	Prov. Cons.	De Provinciis Consularibus
Cic.	Cicero	Leg. Man.	Pro Lege Manilia
Cic.	Cicero	Off.	De Officiis
Cic.	Cicero	Tusc.	Tusculanae Disputationes
Cic.	Cicero	Rep.	De Republica
Cic.	Cicero	Phil.	Orationes Philippicae
Cic.	Cicero	Font.	Pro Fonteio
Cic.	Cicero	Balb.	Pro Balbo
Cic.	Cicero	Leg.	De Legibus
Cic.	Cicero	De. Rep.	De Republica
Cic.	Cicero	Fam.	Ad Familiares
Cic.	Cicero	Att.	Epistulae ad Atticum
Dion. Hal.	Dionysius of Halicarnassus	Ant. Rom.	Antiquitates Romanae
Eutrop.	Eutropius		
Festus	Festus		
Flor.	Florus		
Front.	Frontinus	Strat.	Strategemata
Hdn.	Herodian		
Joseph.	Josephus	B.J.	Bellum Iudicum
Juv.	Juvenal	Sat.	Satires
Livy	Livy		Histories
Lucr.	Lucretius		De Rerum Natura
Onasander	Onasander		The General
Ov.	Ovid	Met.	Metamorphoses
POxy			Papyrus Oxyrhynchus
Plato	Plato	Leg.	Laws
Plato	Plato	Prt.	Protagoras

Abbreviations 159

Abbreviation for author	Author	Abbreviation for work	Name of work
Plaut.	Plautus	*Amp.*	*Amphitruo*
Plaut.	Plautus	*Asin.*	*Asinaria*
Plaut.	Plautus	*Capt.*	*Captivi*
Plaut.	Plautus	*Cist.*	*Cistellaria*
Pliny	Pliny the Younger	*Ep.*	*Epistulae*
Pliny	Pliny the Elder	*Nat.Hist.*	*Naturalis Historia*
Plut.	Plutarch	*Vit. Cic.*	*Life of Cicero*
Plut.	Plutarch	*Vit. Cam.*	*Life of Camillus*
Plut.	Plutarch	*Vit. Marc.*	*Life of Marcellus*
Plut.	Plutarch	*Cat. Mai.*	
Plut.	Plutarch	*Vit. Mar.*	*Life of Marius*
Plut.	Plutarch	*Vit. Pomp.*	*Life of Pompey*
Plut.	Plutarch	*Vit. Crass.*	*Life of Crassus*
Plut.	Plutarch	*Vit. Luc.*	*Life of Lucullus*
Plut.	Plutarch	*Vit. Pomp.*	*Life of Pompey*
Plut.	Plutarch	*Vit. Brut.*	*Life of Brutus*
Plut.	Plutarch	*Vit. Ant.*	*Life of Antony*
Plut.	Plutarch	*Vit. C. Gracch.*	*Life of Caius Gracchus*
Pol.	Polybius		*Histories*
Polyaenus	Polyaenus	*Strat.*	*Stratagems*
Ps-Quint.	Pseudo Quintillian	*Decl. Mai*	*Declamationes Maiores*
Sall.	Sallust	*Cat.*	*Bellum Catalinum*
Sall.	Sallust	*Iug*	*Bellum Jurginthum*
Sall.	Sallust	*H.*	*Histories*
Sen.	Seneca	*Ep.*	*Epistulae*
Sen.	Seneca	*Q.Nat.*	*Quaestiones Naturales*
Suet.	Suetonius	*Tib.*	*Divus Tiberius*
Suet.	Suetonius	*Iul.*	*Divus Iulius*
Tac.	Tacitus	*Hist.*	*Historiae*
Tac.	Tacitus	*Agr.*	*Agricula*
Tac.	Tacitus	*Ann.*	*Annales*
Tac.	Tacitus	*Dial.*	*Dialogus de Oratoribus*

Abbreviation for author	Author	Abbreviation for work	Name of work
Tert.	Tertullian	*DCor.*	*De Corona Militis*
Tert.	Tertullian	*DPal.*	*De Pallio*
Thuc.	Thucydides		
Val. Max.	Valerius Maximus		
Varro		*Ling.*	*De Lingua Latina*
Veg.	Vegetius		*De Re Militari*
Vell. Pat.	Vellius Paterculus		
Xen.	Xenophon	*Eq. Mag.*	*On Horsemanship*

Notes

Introduction
1. Jean-François Lyotard, *La condition postmoderne: rapport sur le savoir* (Paris, 1979), p. 7. For example: Victor Davis Hanson, *Hoplites: The Classical Greek Experience of Battle* (London, 1991), p. 11.
2. Thuc. 1.22; Pol. 1.4.
3. Jeremy Armstrong and Michael P. Fronda, 'Writing about Romans at war', in Jeremy Armstrong and Michael P. Fronda, eds., *Romans at War: Soldiers, Citizens, and Society in the Roman Republic* (London, 2020), pp. 1, 4–5.

Chapter 1
1. Common works in English, French and German from only the last thirty years include: Simon Elliot, *Romans at War: the Roman Military in the Republic and Empire* (Oxford, 2020), Conor Whately, *An Introduction to the Roman Military: From Marius (100 BCE) to Theodosius II (450 CE)* (Hoboken, 2021), David J. Breeze, *The Roman Army* (London, 2016), Yann Le Bohec, ed., *The Encyclopedia of the Roman Army* (Chichester, West Sussex, 2015), Alexander Rudow, *Die römische Armee: Organisation, Ausrüstung, Eroberungen* (Rheinbach, 2015), Catherine Wolff, *L' armée romaine: une armée modèle* (Paris, 2012), Pierre Streit, *L'armée romaine* (Gollion, 2012), Nigel Rodgers, *Die römische Armee: die Legionen der antiken Weltmacht und ihre Feldzüge* (Tosa, 2011), Pat Southern, *The Roman Army: a Social and Institutional History* (Oxford, 2007), Arthur Keaveney, *The Army in the Roman Revolution* (London, 2007), Graham Sumner, *Die römische Armee: Bewaffnung und Ausrüstung* (Stuttgart, 2007), Paul Erdkamp, ed., *A Companion to the Roman Army* (Oxford, 2007), Yann Le Bohec, *L' armée romaine sous le bas-empire* (Paris, 2006), Nigel Rodgers, *The Roman Army: Legions, Wars and Campaigns: A Military History of the World's First Superpower from the Rise of the Republic and the Might of the Empire to the Fall of the West* (London, 2005), Michael Simkins and Ronald Embleton, *Die römische Armee: von Caesar bis Constantin (44 v. Chr.-333 n. Chr.)* (Sankt Augustin, 2005), Adrian Keith Goldsworthy, *The Complete Roman Army* (London, 2003), G. Webster, *The Roman Imperial Army* (Norman, 1998), Adrian Keith Goldsworthy, *The Roman Army at War: 100 BC-AD 200* (Oxford, 1996) and Y. Le Bohec, *The Imperial Roman Army* (London, 1994).
2. Goldsworthy, *The Roman Army at War: 100 BC-AD 200*, p. 2.
3. On military manuals, see Conor Whately, 'Military manuals from Aeneas Tacitus to Maurice. Origins, scholarship, genre, audience and history', in James T. Chlup and Conor Whately, eds., *Greek and Roman Military Manuals: Genre and History* (London, 2021).
4. Pol 1.1.5–6, Praef 1.2. See A.M. Eckstein, *Mediterranean Anarchy, Interstate War, and the Rise of Rome* (Berkeley, 2009), p. 100f, 104. Polybius appears to have believed that Macedonian soldiers set the benchmark for military ferocity: *Ibid.*, pp. 202–3.

5. F. W. Walbank, *Polybius, Rome and the Hellenistic World: Essays and Reflections* (Cambridge, 2002), p. 278.
6. Pol 6.25.11.
7. Y. Le Bohec, 'Roman Wars and Armies in Livy', in Bernard Mineo, ed., *A Companion to Livy* (Chichester, 2015), p. 114.
8. Livy Praef. 9, translation: Livy, *Livy with an English Translation in Fourteen Volumes* (Cambridge, Mass., 1982).
9. P.G. Walsh, *Livy* (Cambridge, 1961), p. 83; Jane D. Chaplin, *Livy's Exemplary History* (Oxford, 2000), Introduction.
10. T.J. Luce, *Livy: The Composition of his History* (Princeton, 1977), p. 230f.
11. A. Feldheer, *Spectacle and Society in Livy's History* (Berkeley, 1998), Ch 3: '*Devotio*'. Livy 10.29.
12. Feldheer, *Spectacle and Society in Livy's History*, Ch. 3: 'Camillus the Historian'.
13. Sam Koon, *Infantry Combat in Livy's Battle Narratives* (Oxford, 2010), pp. 82–3.
14. Livy is read by Le Bohec to satisfy the interest in the Romans' tactics and organization although with some difficulty: Le Bohec, 'Roman Wars and Armies in Livy', p. 114.
15. C.T. Allmand, *The De Re Militari of Vegetius: The Reception, Transmission and Legacy of a Roman Text in the Middle Ages* (Cambridge, 2011), p. 1f.
16. George T. Dennis, *The Taktika of Leo VI* (Washington, 2014), p. ix.
17. John F. Haldon, *Warfare, State and Society in the Byzantine world 565–1204* (London, 1999), p. 5.
18. Imperator Mauricius, *Maurice's Strategikon: Handbook of Byzantine Military Strategy* (Philadelphia, 1984), pp. xvi–xvii. Ernst Gamillscheg and George T. Dennis, eds., *Das Strategikon des Maurikios* (Wien, 1981), p. xii.B.8, xii.B.Pr.
19. Haldon, *Warfare, State and Society in the Byzantine World 565–1204*, pp. 5, 7.
20. For the particular interests of the Byzantines, see: Meredith L.D. Riedel, '"God has sent thunder". Ideological distinctives of middle Byzantine military manuals', in James T. Chlup and Conor Whately, eds., *Greek and Roman Military Manuals: Genre and History* (London, 2021).
21. Allmand, *The De Re Militari of Vegetius: The Reception, Transmission and Legacy of a Roman Text in the Middle Ages*, p. 329f. Lionel Kenneth Carley, 'The Anglo-Norman Vegetius: a Thirteenth Century Translation of the "De re militari"', (PhD thesis, Nothingham, 1962), p. 8.
22. Elizabeth L. Eisenstein, *The Printing Revolution in Early Modern Europe* (Cambridge, 2013).
23. A. Momigliano, 'Ancient History and the Antiquarian', *Journal of the Warburg and Courtauld Institutes*, 13 (1950), pp. 289, 291.
24. Azar Gat, *The Origins of Military Thought from the Enlightenment to Clausewitz* (Oxford, 1989), p. 7. Angelo Mazzocco and Marc Laureys, *A New Sense of the Past: The Scholarship of Biondo Flavio (1392–1463)* (Leuven, 2016), p. 14. Biondo devoted Books VI-VII of his work *Roma Triumphans* to Rome's armies.
25. Justus Lipsius, *Iusti Lipsi de militia Romana libri quinque, commentarius ad Polybium.* (Antveria, 1598).
26. K. A. E. Enenkel, Koen Ottenheym and Alexander C. Thomson, *Ambitious Antiquities, Famous Forebears: Constructions of a Glorious Past in the Early Modern Netherlands and in Europe* (Leiden, 2019), p. 158f.
27. Gat, *The Origins of Military Thought from the Enlightenment to Clausewitz*, p. 2.

28. N. Machiavelli, *The Art of War* (New York, 1965), pp. xxix–xxx.
29. Niccolo Machiavelli, *Discourses on the First Decade of Titus Livius* (London, 1883).
30. *Ibid.*, p. xvii.
31. Niccolo Machiavelli, *Discorsi di Nicolo Machiavelli fiorentino, sopra la prima deca di Tito Livio* (Venegia, 1554), p. I, 4; III, 36
32. Machiavelli, *Discourses on the First Decade of Titus Livius*, p. 23.
33. *Ibid.*, p. 45–49.
34. *Ibid.*, p. 20–21.
35. Gat, *The Origins of Military Thought from the Enlightenment to Clausewitz*, p. Ch. 1.
36. Raimondo Montecuccoli, *Mémoires de Montecucculi Generalaissime des Troupes de l'Emperor, divisez en trois livres. 1. De l'art militaire en général. 2. De la guerre contre le Turc. 3. Relation de la Campagne de 1664.* (Strasbourg, 1735), p. 24.
37. *Ibid.*, p. I.ii.xiii.
38. *Ibid.*; I.ii.xv; I.ii.xxi; I.iii.xlvii; I.iii.xlviii
39. Montecuccoli, *Mémoires de Montecucculi Generalaissime des Troupes de l'Emperor, divisez en trois livres. 1. De l'art militaire en général. 2. De la guerre contre le Turc. 3. Relation de la Campagne de 1664*, p. I.ii.xiii.
40. *Ibid.*, p. I.ii.xv.iv.
41. *Ibid.*, p. I.ii,xv.ii.
42. Jean Charles Chevalier de. Folard, *Mémoires pour servir à l'histoire de Monsieur le Chevalier de Folard.* (Ratisbonne, 1753), p. 85.
43. *Ibid.*, p. 103.
44. C. Duffy, *The Military Experience in the Age of Reason* (London, 1987), p. 198f. De Folard, *Mémoires pour servir à l'histoire de Monsieur le Chevalier de Folard*, pp. 104–5. Gat, *The Origins of Military Thought from the Enlightenment to Clausewitz*, pp. 8, 38.
45. M. De Saxe, *Mes Rêveries* (Amsterdam, 1757), pp. 32–3.
46. *Ibid.*, pp. 22, 133. De Saxe may have gained this impression from Vegetius' reference to the need to train recruits in the use of the *gradus militaris:* Publius Flavius Renatus Vegetius (M. D. Reeve, ed), *Vegetius Epitoma Rei Militaris*, (Oxford, 2004), p. 1.9. The cadence step did not appear in European armies until the late seventeenth century and probably arose from the evolutions for the loading and reloading of firearms: J. Keegan, *A History of Warfare* (London, 1993), pp. 341–2. See also W.H. McNeill, *The Pursuit of Power* (Oxford, 1983), pp. 128–9. Maurice of Nassau in promoting drill was inspired by the Romans.
47. De Saxe, *Mes Rêveries*, pp. 39–40.
48. B.P. Hughes, *Firepower* (London, 1974), Ch. 5. See also George F. Nafziger, *Imperial Bayonets: Tactics of the Napoleonic Battery, Battalion, and Brigade as Found in Contemporary Regulations* (London, 1996), p. 31f.
49. De Saxe, *Mes Rêveries*, p. 47f.
50. *Ibid.*, p. 121.
51. *Ibid.*, p. 108.
52. *Ibid.*, p. 40.
53. Gat, *The Origins of Military Thought from the Enlightenment to Clausewitz*, p. 26.
54. *Ibid.*, p. 8.
55. *Ibid.*, p. 28.
56. M. Joly de Maizeroy, *Cours De Tactique Théorique, Pratique, Et Historique, Qui applique les exemples aux préceptes, développe les maximes des plus habiles Généraux, & rapporte les*

faits les plus intéressans & les plus utiles; avec les descriptions de plusieurs batailles anciennes (vol. 1, Paris, 1766), I, p. ii.
57. *Ibid.*, p. iii.
58. *Ibid.*, pp. x, xi.
59. Charles Guischardt, *Mémoires militaires sur les Grecs et les Romains, où l'on a fidélement retabli sur le texte de Polybe et des tacticiens Grecs et Latins* (vol. 1, La Haie, 1758), I, p. Discours Preliminaire.
60. *Ibid.*, p. 1f.
61. Karl Gottlieb Guischardt, *Principes de l'art militaire: extraits des meilleurs ouvrages des Anciens* (vol. 1, Berlin, 1763), p. 57.
62. *Ibid.*, p. 58f.
63. *Ibid.*, p. 113.
64. *Ibid.*, p. 116.
65. *Ibid.*, p. 3.
66. Jacques-Antoine-Hippolyte Comte de Guibert, *Oeuvres Militaries de Guibert* (vol. 1, Paris, 1803), I, pp. 81–84.
67. Gat, *The Origins of Military Thought from the Enlightenment to Clausewitz*, p. 45.
68. de Guibert, *Oeuvres militaries de Guibert*, pp. 90, 130.
69. Gat, *The Origins of Military Thought from the Enlightenment to Clausewitz*, p. 39f.
70. *Ibid.*, pp. 147–8.
71. *Ibid.*, p. 59.
72. *Ibid.*, p. 62.
73. Georg Heinrich von Berenhorst, *Betrachtungen über die Kriegskunst, über ihre Fortschritte, ihre Widersprüche und ihre Zuverläßigkeit. Auch für Laien verständlich, wenn sie nur Geschichte wissen* (Leipzig, 1797–1799).
74. Gat, *The Origins of Military Thought from the Enlightenment to Clausewitz*, pp. 153–4.
75. For example: Baron De Jomini, trans. Maj. W.O. Winship and Lieut. E.E. McLean, *Summary of the Art of War or, A New Analytical Compend of the Principal Combinations of Strategy, of Grand Tactics and of Military Policy* (New York, 1854); Baron De Jomini, *The Art of War by Baron De Jomini* (Philadelphia, 1862), pp. 1–2; Baron Jomini, trans. Mendell, Capt. G.H. and Craighill, Lieut. W.P., *Treatise on Grand Military Operations or A Critical and Military History of the Wars of Frederick the Great as contrasted with the Modern System* (New York, 1865).
76. A.H. Baron de Jomini, *Traité des grandes opérations militaires, contenant l'histoire critique des campagnes de Frédéric II, comparées à celles de l'empereur Napoléon: avec un recueil des principes généraux de l'art de la guerre* (vol. I, Paris, 1811), I, p. 258, fn 1. A.H. Baron de Jomini, *Traité des grandes opérations militaires, contenant l'histoire critique des campagnes de Frédéric II, comparées à celles de l'empereur Napoléon : avec un recueil des principes généraux de l'art de la guerre* (vol. III, Paris, 1811), pp. 158–9, 164, 198, 235, 280 fn 1, 340. Baron De Jomini, *Précis de l'art de la guerre, ou nouveau tableau analytique des principales combinations de la stratégie, de la grande tactique et de la politique militaire* (vol. Partie I, Paris, 1837a), pp. 19, 49, 179. Baron De Jomini, *Précis de l'art de la guerre, ou nouveau tableau analytique des principales combinations de la stratégie, de la grande tactique et de la politique militaire* (vol. Partie II, Paris, 1838), p. 138. Baron De Jomini, *Précis de l'art de la guerre, ou nouveau tableau analytique des principales combinations de la stratégie, de la grande tactique et de la politique militaire* (vol. II, Bruxelles, 1840), p. 40.

77. A.H. Baron de Jomini, *Traité des grandes opérations militaires, contenant l'histoire critique des campagnes de Frédéric II, comparées à celles de l'empereur Napoléon avec un recueil des principes généraux de l'art de la guerre* (vol. II, Paris, 1811), p. 198.
78. De Jomini, *The Art of War by Baron De Jomini*, pp. 22, 48.
79. De Jomini, *Précis de l'art de la guerre, ou nouveau tableau analytique des principales combinations de la stratégie, de la grande tactique et de la politique militaire*, p. 97.
80. *Ibid.*, p. 122.
81. De Jomini, *Précis de l'art de la guerre, ou nouveau tableau analytique des principales combinations de la stratégie, de la grande tactique et de la politique militaire*, p. 200.
82. *Ibid.*, pp. 47, 107.
83. Gat, *The Origins of Military Thought from the Enlightenment to Clausewitz*, p. 9.
84. C. von Decker, *De la tactique des trois armes, infanterie, cavalerie, artillerie, isolées et réunies dans l'esprit de la nouvelle guerre cours fait à l'École militaire de Berlin par C. de Decker: Contenant la tactique de chaque arme isolée* (vol. 1, Brussels, 1836), p. 33.
85. Jean Thomas Rocquancourt, *Cours complet d'art et d'histoire militaires : ouvrage dogmatique, littéraire et philosophique à l'usage des élèves de l'école royale spéciale militaire* (vol. 1, Paris, 1840), p. 292, 298–9, 302, 319, 437–8.
86. *Ibid.*, p. 24.
87. *Ibid.*, p. 37.
88. *Ibid.*, p. 49.
89. *Ibid.*, p. 51.
90. *Ibid.*, pp. 52f, 56.
91. *Ibid.*, pp. 69, 332.
92. Colonel Marie Henri François Carrion-Nisas, *Essai sur l'histoire générale de l'art militaire, des son origine, de ses progrès et des ses révolutions, depuis la première formation des sociétés européennes jusqu'à nos jours, orné de quatorze places* (vol. 2, Paris, 1824), II, p. 64.
93. *Ibid.*, p. 103.
94. *Ibid.*, p. 161.
95. Wilhelm Rüstow, *Der Krieg und seine Mittel; eine allgemein fassliche Darstellung der ganzen Kriegskunst.* (Leipzig, 1856).
96. Wilhelm Rüstow, *Heerwesen und Kriegführung C. Julius Cäsars* (Nordhausen, 1862), p. vi.
97. Rüstow, *Der Krieg und seine Mittel; eine allgemein fassliche Darstellung der ganzen Kriegskunst*, p. 466.
98. H.P. Judson, *Caesar's Army; a Study of the Military Art of the Romans in the Last Days of the Republic* (New York, 1888).
99. Nicolas Édouard de La Barre Duparcq, *Éléments d'art et d'histoire militaire* (Paris, 1858), pp. 137, 140f, 144, 15.
100. Pol. 6.25.11
101. de La Barre Duparcq, *Éléments d'art et d'histoire militaire*, p. 139, fn 1.
102. Franz Fröhlich, *Das Kriegswesen Cäsars* (vol. 1, Zürich, 1890), I, pp. iii–iv.
103. Theodor Steinwender, *Die Marschordnung des römischen Heeres zur Zeit der Manipularstellung* (Danzig, 1907), p. 42.
104. Col. Ardant du Picq, trans Col. John N. Cotton and Maj. Robert C. Cotton, *Battle Studies Ancient and Modern* (New York, 1921).
105. A. Du Picq, *Études sur le Combat* (Paris, 1880), p. vi.

106. Momigliano, 'Ancient History and the Antiquarian', p. 292.
107. *Ibid.*, p. 293f.
108. *Ibid.*, p. 307.
109. For example: Albert Schwegler, *Römische Geschichte im Zeitalter der Könige* (vol 1, Tübingen, 1869)., Albert Schwegler, *Römische Geschichte im Zeitalter des Kampfs der Stände* (vol. 2, Tübingen, 1872). and Albert Schwegler, *Römische Geschichte von gallischen Brande Roms bis zum ersten Samniter Kreige* (vol. 4, Tübingen, 1873).
110. For example: Joachim Marquardt and Theodor Mommsen, *Handbuch der Römischen Alterthümer, Römische Staatsverwaltung von Joachim Marquardt, Zweiter Band* (vol. 5, Leipzig, 1884), V, pp. v–vi.
111. Karl Joachim Marquardt and Theodor Mommsen, *Handbuch der römischen Alterthümer, Staatsverwaltung von J. Marquardt II* (vol. 5, Leipzig, 1876), V, p. 309f.
112. F. Cadiou, *L'armée imaginaire: les soldats prolétaires dans les légions Romaines au dernier siécle de la République* (Paris, 2018), pp. 61–2.
113. Rocquancourt, *Cours complet d'art et d'histoire militaires : ouvrage dogmatique, littéraire et philosophique à l'usage des élèves de l'école royale spéciale militaire*, p. 104.
114. *Ibid.*, p. 423.
115. Clovis Lamarre, *De la milice romaine depuis la fondation de Rome jusqu'à Constantin* (Paris, 1863).
116. Ludwig Lindenschmit, *Tracht und Bewaffnung des römischen Heeres während der Kaiserzeit* (Braunschweig, 1882), p. 2.
117. Conrad Cichorius, *Die reliefs der Traianssäule.* (2 vols, Berlin, 1896).
118. Carl Adolf Löhr, *Ueber die Taktik und das Kriegswesen der Griechen und Römer* (Kempten, 1825), pp. iii–iv.
119. Felix Joseph Lipowsky, *Des Flavius Vegetius Renatus fünf Bücher über die Kriegswissenschaft und Kriegskunst der Römer* (Sulzbach, 1827), p. viii.
120. Jules de la Chauvelays, *L'Art militaire chez les Romains. Nouvelles observations critiques pour faire suite à celles du Chevalier Folard et du Colonel Guischardt. Avec une lettre du Général Davout, Duc d'Auerstædt.* (Paris, 1884), p. i.
121. A. Harkness, *The Military System of the Romans* (New York, 1887), pp. xxvii, xxix–xxxi, xxxiii, xxxv, xxxix, xlv–xlvi, l–li, lx, xlii, lxiv, lxvii–lxviii, lxxf.
122. A. von Domaszewski, *Die Rangordnung des Römischen Heeres* (Bonn, 1908), p. 192.
123. *Ibid.*, p. 193.
124. E. Ritterling, 'Legio. Bestand, Verteilung und kriegerische Betätigung der Legionen des stehenden Heeres von Augustus bis Diocletian', *Paulys Realencyclopädie der classischen Altertumswissenschaft*, XII, 1 (1924), col. 1211–1328; W. Kubitschek, 'Legio. Republikanische Zeit', *Paulys Realencyclopädie der classischen Altertumswissenschaft*, XII, 1 (1924), col. 1186–1210.
125. René Cagnat, *L'armée romaine d'Afrique et l'occupation militaire de l'Afrique sous les empereurs* (Paris, 1913).
126. G.L. Cheesman, *The Auxilia of the Roman Army* (Oxford 1914), p. 13.
127. F. Haverfield, 'Obituary: Leonard Cheesman', *Journal of Roman Studies*, 5 (1915), p. 147.
128. H.M.D. Parker, *The Roman Legions* (London, 1928), p. 6.
129. Eric Birley, 'Hadrian's Wall', *The Antiquaries Journal*, XI (1931), p. 62.
130. Rüstow, *Der Krieg und seine Mittel; eine allgemein fassliche Darstellung der ganzen Kriegskunst*, p. 3f.

131. Rüstow, *Heerwesen und Kriegführung C. Julius Cäsars*, p. 10.
132. Karl Gustav von Berneck, *Geschichte der Kriegskunst für Militairakademien und Offiziere aller Grade*. (Berlin, 1867), p. iii.
133. J. von Hardegg, *Anleitung zum Studium der Kriegsgeschichte*. (vol. 1, Darmstadt, 1868), I, pp. 6, 224, 235f, 240–1, 243–4, 246.
134. F.A. Paris, *Traite de tactique appliquée : élaboré d'après le programme prescrit pour les écoles royales de guerre allemandes* (Paris, 1873), p. 5, 315.
135. Max Jähns, *Handbuch einer Geschichte des Kriegswesens von der Urzeit bis zur Renaissance: technischer Theil: Bewaffnung, Kampfweise, Befestigung, Belagerung, Seewesen* (Leipzig, 1880), p. ix.
136. Albert Wilms, *Die Schlacht bei Cannae* (Hamburg, 1895), p. 3.
137. *Ibid.*, p. 1.
138. Armstrong and Fronda, 'Writing about Romans at war', pp. 4–5.
139. A. Bucholz, *Hans Delbrück and the German Military Establishment: War Images in Conflict* (Iowa City, 1985), p. xi.
140. *Ibid.*, p. 10.
141. *Ibid.*, p. 27.
142. Fernando Quesada Sanz, 'Not so different: individual fighting techniques and battle tactics of Roman and Iberian armies within the framework of warfare in the Hellenistic Age', in P. François, P. Moret and S. Péré-Noguès, eds., *L'Hellénisation en méditerranée occidentale au temps des guerres puniques. Actes du Colloque International de Toulouse, 31 mars-2 avril 2005 Pallas 70* (2006), p. 245.
143. Bucholz, *Hans Delbrück and the German Military Establishment: War Images in Conflict*, p. 28f.
144. For example: Hans Delbrück, *Warfare in Antiquity: History of the Art of War* (vol. 1, Lincoln, 1990), I, p. 466.
145. Bucholz, *Hans Delbrück and the German Military Establishment: War Images in Conflict*, p. 31f.
146. H. Delbrück, *Geschichte der Kriegskunst im Rahmen der Politischen Geschichte: das Altertum* (vol. I, Berlin, 1920), I, p. 291–2.
147. *Ibid.*, p. 448. On the change from manipular to cohortal organisation, see Cadiou, *passim*.
148. *Ibid.*, p. 389.
149. *Ibid.*, p. 462.
150. Georg Veith, *Geschichte der Feldzüge C. Julius Caesars* (Wien, 1906), p. v.
151. *Ibid.*, p. viii.
152. J. Kromayer and G. Veith, *Heerwesen und Kriegsführung der Griechen und Römer* (München, 1928), p. v.
153. *Ibid.*, p. vi.
154. Bucholz, *Hans Delbrück and the German Military Establishment: War Images in Conflict*, p. 66f.
155. Alfred von Schlieffen and Hugo Friedrich von Freytag-Loringhoven, *Cannae mit einer Auswahl von Aufsätzen und Reden des Feldmarschalls sowie einer Einführung und Lebensbeschreibung von General der Infanterie Freiherr von Freytag-Loringhoven* (Berlin, 1925).
156. General Field Marshal Alfred von Schlieffen, trans. Anon., *Cannae* (Fort Leavenworth, 1931).

157. Hermann von François, *Tannenberg; Das Cannae des Weltkrieges in Wort und Bild* (Berlin, 1926).
158. Hanson, *Hoplites: The Classical Greek Experience of Battle*, p. 10.
159. E.N. Luttwak, *The Grand Strategy of the Roman Empire* (Baltimore, 1976), p. 76. E.L. Wheeler, 'Methodological Limits and the Mirage of Roman Strategy: Part 1', *The Journal of Military History*, 57 (1993), *passim*.
160. Luttwak, *The Grand Strategy of the Roman Empire*. His views, although supported by Wheeler: Wheeler, 'Methodological Limits and the Mirage of Roman Strategy: Part 1 passim, James Lacey, 'The grand strategy of the Roman Empire', in Richard Hart Sinnreich and Williamson Murray, eds., *Successful Strategies: Triumphing in War and Peace from Antiquity to the Present* (Cambridge, 2014), pp. 38–64 and Arthur Ferrill, *Roman Imperial Grand Strategy* (Lanham, 1991), have not been accepted by others, notably Isaacs: B. Isaac, *The Limits of Empire: The Roman Army in the East* (Oxford, 1990) and by Whittaker: C.R. Whittaker, *The Frontiers of the Roman Empire: A Social and Economic Study* (Baltimore, 1994) and C.R. Whittaker, *Rome and its Frontiers: The Dynamics of Empire* (London, 2004).

Chapter 2

1. See Giuseppe Caforio, ed., *Handbook of the Sociology of the Military* (New York, 2006), p. 4f.
2. Eugene Kameka, 'Political Nationalism – the Evolution of the Idea', in Eugene Kamenka, ed., *Nationalism: The Nature of an Idea* (Canberra, 1973), pp. 7–8. Azar. Gat, *War in Human Civilization* (Oxford, 2006), pp. 498f, 540. Azar Gat and Alexander Yakobson, *Nations: The Long History and Deep Roots of Political Ethnicity and Nationalism* (Cambridge, 2013), p. 143.
3. Kameka, 'Political Nationalism – the Evolution of the Idea', p. 9f.
4. Anthony W. Marx, *Faith in Nation: Exclusionary Origins of Nationalism* (Oxford, 2003), pp. 29, 204f. Marx argues that nationalism has sectarian origins and is a construction of reality to suit the needs of secular rulers. Caspar Hirschi agrees that nationalism has confessional roots while having a distinctly humanist offshoot in nationalism: Caspar Hirschi, *The Origins of Nationalism: An Alternative History from Ancient Rome to Early Modern Germany* (Cambridge, 2012), pp. 203f, 209f.
5. Not all military historians have appreciated this: Everett Thomas Dague, *Napoleon and the First Empire's Ministries of War and Military Administration: The Construction of a Military Bureaucracy* (Lewiston, N.Y., 2006), p. 19.
6. Isabel V. Hull, *Absolute Destruction: Military Culture and the Practices of War in Imperial Germany* (London, 2005), p. 100f.
7. Dague, *Napoleon and the First Empire's Ministries of War and Military Administration: The Construction of a Military Bureaucracy*, p. 24f.
8. *Ibid.*, Ch. 11.
9. George Q. Flynn, *Conscription and Democracy: The Draft in France, Great Britain, and the United States* (London, 2002), pp. 3, 5.
10. Christopher M. Clark, *Iron Kingdom: The Rise and Downfall of Prussia, 1600–1947* (London, 2007), p. 558.
11. *Ibid.*, pp. 600–603.
12. 'Killing for the State, Dying for the Nations: An Introductory Essay on the Life Cycle of Conscription onto Europe's Armed Forces', in Lars. Mjøset, and Stephen.

Van Holde, eds., *The Comparative Study of Conscription in the Armed Forces* (Oxford, 2002), p. 47.
13. Flynn, *Conscription and Democracy: The Draft in France, Great Britain, and the United States*, p. 17.
14. *Ibid.*, pp. 18–21.
15. *Ibid.*, p. 4.
16. For example: R. Alston, *Soldier and Society in Roman Egypt: A Social History* (London, 1995), pp. 4–5; A. Santosuosso, *Storming the Heavens* (London, 2004), pp. 104–6, also 'The New Army', 51: 'The army of Caesar's Gallic wars was an array of men that could be called a national army', citing Harmond in Brisson, J.-P. *Problèmes de la guerre à Rome* (Paris, 1969), p. 71. One modern work on the Roman army is explicitly subtitled 'A social and institutional history': Southern, *The Roman Army: A Social and Institutional History*.
17. S. James, 'Writing the Legions: The Development and Future of Roman Military Studies in Britain', *Archaeology Journal*, 159 (2002), p. 38. The examples of the use of *exercitus* were taken from a search of the digital version of the *Thesaurus Linguae Latinae*.
18. Veg. *Mil.* 2.9, 3.1. See also *Cod Iust.* 3.2.2.
19. Vell. Pat. 2.16; Livy, 1.12, uses *Romanus exercitus* to distinguish the Roman force from that of the Sabines: *Tenuere tamen arcem Sabini atque inde postero die cum Romanus exercitus instructus quod inter Palatinum Capitolinumque collem campi est complesset non prius decenderunt in aequum quam ira et cupiditate recuperandae arcis stimulante animos in adversum Romani subiere*. Further examples can be found in 43.6 and Tac. *Hist.* 4.57. Sal. *Cat.* 61. Sallust refers to the army of the Roman people: *neque tamen exercitus populi Romani laetam aut incruentam victoriam adeptus erat. Exercitus noster:* Cic. *Prov. Cons.* 5. See also Vell. Pat. 2.111 in reference to Tiberius and Tacitus' description of a Roman force as it advanced: Tac. *Ann.* 2.16. Vegetius contrasts the Roman army with enemy forces: Veg. *Mil.* 3.9. For armies from places: For example: Livy 21.40; Sall. *H.* 2.47.7; Tac. *Agr.* 41; Veg. *Mil.* 3.10; Deutsche Akademie Der Wissenschaften Zu Berlin, 'Corpus Inscriptionum Latinarum', (Berlin, 1893 -1998), pp. 6, 1450. *L'Année épigraphique* (Paris, 1888), 18c, 903, 79, 1905, 143, 1905, 144, 1908, 237. For Hadrian's series of coins referring to armies in different provinces, see C. Foss, *Roman Historical Coins* (London, 1990), p. 115f.
20. See Livy 10.18; Caes. *BCiv.* 1.49; Val. Max. 7.6; Tac. *Ann.* 2.15; Livy 26.42.2.
21. Foss, *Roman Historical Coins*, p. 77.
22. *Ibid.*, pp. C360, 77, 95.
23. Cic. *de Or.* 1.210; Sall. *Iug* 45.2; Vell. 2.47.2; Liv. 44.39.
24. Sall. *Iug* 46.6.
25. An interesting example is found in the recently discovered inscription: *Senatus Consultum de Cn. Pisone Patre*, 55–6 where troops (*milites*) are described as *Pisoniani* and *Caesariani*: D.S. Potter and C. Damon, 'The Senatus Consultum de Cn Pisone Patre', *American Journal of Philology*, 120 (1999), pp. 13–42.
26. For other examples, see Cic. *Leg. Man.* 10.28; Cic. *Off.* 3.26.97; Caes. *B. Gall.* 6.18, 6.14; Sall. *Iug.* 63.2; Sall. *Cat.* 7.4; Vell. 2.5.1; Juv. *Sat.* 16.2, 16.53.
27. Cic. *Tusc.* 5.19.55; Cic. *De Or.* 3.33.134; Liv. 7.32.
28. Juv. *Sat.* 16.2. See also 16.53.
29. The contrast between *militia Caesaris* and *militia Christi* is the focus of Tert. *D. Cor* 11 and 12 is implied particularly in 11.3, 12. 5. On *militia Christi*, see Adolf von

Harnack, *Militia Christi: die christliche Religion und der Soldatenstand in den ersten drei Jahrhunderten* (Tübingen, 1905)., John Helgeland, 'Christians and the Roman Army, A.D. 173–337', *Church History: Studies in Christianity and Culture*, XLIII (1974), pp. 149–163.
30. S. Riccobono, G. Baviera, C. Ferrini, G. Furlani, V. Arangio-Ruiz, 'Fontes iuris romani antejustiniani', (3 Vols, Florence, 1940), III, pp. 1, 93, 8–10.
31. Amm. Marc. 31.11.
32. Veg. *Mil.* prooemium 2.
33. C. Kelly, *Ruling the Later Roman Empire* (Cambridge, Mass., 2004), p. 20.
34. Smith finds 'standing armies': R. E. Smith, *Service in the Post-Marian Roman Army* (Manchester, 1958), Ch II, 28. Keppie argues that Smith's distinction between 'emergency armies' and 'standing armies' is unnecessary but accepts that oaths were sworn to each commander: Lawrence Keppie, *The Making of the Roman Army* (London, 1984), pp. 77–8.
35. Cass. Dio 52.27.1–2. It ought to be noted that Cary, the Loeb translator, translates στρατιώτας ἀθανάτους as 'standing army': Cassius Dio, *Dio's Roman History, in Nine Volumes* (Cambridge, Mass., 1968).; F. Millar, *A Study of Cassius Dio* (Oxford, 1964), p. 109.
36. M. Reinhold, 'In praise of Cassius Dio', *L'Antiquité Classique*, 55 (1986), pp. 214–15. P.M. Swan, 'How Cassius Dio composed his Augustan books: four studies.', *ANRW*, Principät II (1997), pp. 2524–5.
37. Hdn. 2.11.4–5.
38. Josephus, *BJ* 3.30f.
39. For the changes which occurred during Augustus' principate, see Fred K. Drogula, *Commanders and Command in the Roman Republic and Early Empire* (Chapel Hill, 2015), Ch 7.

Chapter 3

1. The earliest instance of the use of reform to describe change in the Roman armed forces that I have found is in a translation of a French work of 1857: E. de la Barrie Duparcq, *Duparcq's Military Art and History* (New York, 1983), p. 22–3. On the Marian army reforms, Bell credits Marquardt with the theory of the Marian Army Reforms: M.J.V. Bell, 'Tactical Reform in the Roman Republican Army', *Historia*, 14 (1965), p. 404.
2. Hooke, Nathaniel, *The Roman History from the Building of Rome to the Ruin of the Commonwealth* (vol. II, Dublin, 1759), pp. 228, 478, 493–5, 519 and in Hooke, Nathaniel, *The Roman History from the Building of Rome to the Ruin of the Commonwealth* (vol. III, Dublin, 1759), pp. 211, 250.
3. Edward Gibbon, *The Decline and Fall of the Roman Empire* (London, 1995), I, pp. 9, 24.
4. For example: Goldsmith, *Goldsmith's Roman History: Abridged by Himself, for the Use of Schools* (Poughkeepsie, 1816), pp. 80, 212, 285.
5. Schmitz, Leonhard, *Lectures on the History of Rome: From the Earliest Times to the Fall of the Western Empire by R.G. Niebuhr* (vol. I, London, 1850), I, pp. 177, 235, 241, 279, 345–6, 348, 390. Smith, W. G., *Dictionary of Greek and Roman Biography and Mythology* (vol. 3, London, 1849), III, pp. 657, 754, 807, 942, 1004, 1243, 1247, 1348. See also F.W. Newman, *Regal Rome: An Introduction to Roman History* (London, 1852), pp.

134, 144. and G.C. Lewis, *An Inquiry into the Credibility of the Early Roman History* (vol. II, London, 1855), II, pp. 154, 542.
6. T. Arnold, *The History of Rome* (vol. I, New York, 1846), I, pp. 86, 128, 235.
7. Liddell, Henry G., *A History of Rome from the Earliest Times to the Establishment of the Empire* (vol. I, London, 1855), I, pp. ix, xi, 43, 124, 198, 193, 417, 431; C. Merivale, *History of the Romans Under the Empire* (vol. I, New York, 1863), I, pp. 13, 17, 59, 126–7, 163, 169, 346, 362, 435.
8. R.F. Leighton, *A History of Rome* (New York, 1889), pp. 4, 44, 116; H.F. Pelham, *Outlines of Roman History* (London, 1893), pp. 6, 195, 207. W.W. Capes, *Roman History: The Early Empire* (London, 1897), pp. 7, 14, 214.
9. Philip Van Ness Myers, *A History of Rome* (London, 1904), pp. 27, 43, 44. F.F. Abbott, *A Short History of Rome* (New York, 1906), pp. 64, 114, 148. H. Webster, *Ancient History* (Boston, 1913), pp. 173, 396, 399. Tenney Frank, *A History of Rome* (New York, 1923), pp. 40, 153.
10. Search on Google Books: http://books.google.com/books?lr=&as_brr=0&id =zwIfAAAAMAAJ&dq=reform+%22history+of+rome%22+date%3A1925- 1950&q=reform&pgis=1#search (accessed 20 Dec. 2007).
11. M. Cary, *A History of Rome down to the Reign of Constantine* (London, 1938), p. 80.
12. *Ibid.*, p. 86.
13. *Ibid.*, p. 240.
14. *Ibid.*, p. 244.
15. *Ibid.*, p. 250.
16. *Ibid.*, p. 253.
17. *Ibid.*, p. 282.
18. For example: Frank, *A History of Rome*, pp. 40, 153; M.I. Rostovtzeff, *A History of the Ancient World* (vol. II, London, 1945), II, pp. 49, 106, 107.
19. F.W. Walbank, A.E. Astin, M.W. Frederiksen and R.M. Ogilvie, eds., *The Cambridge Ancient History Volume VII Part 2: The Rise of Rome to 220 BC.* (Cambridge, 1989), pp. 91, 103, 106, 118, 204, 216, 223, 236, 238, 240, 242, 301, 327, 334, 336, 341–2, 344, 432, 437, 439, 440–41, 443, 611, 613, 622.
20. A.E. Astin, F.W. Walbank, M.W. Frederiksen and R.M. Ogilvie, eds., *The Cambridge Ancient History Volume 8: The Cambridge Ancient History: Rome and the Mediterranean to 133 BC* (Cambridge, 1989), pp. 75, 186, 195, 218, 227, 235, 348, 566, 601, 609.
21. J.A. Crook, Andrew Lintott and Elizabeth Rawson, *The Cambridge Ancient History Volume IX: The Last Age of the Roman Republic, 146–43 BC* (1994), pp. 5, 50, 80–81, 173, 190, 200, 202, 225, 227, 299, 302, 337, 349, 353, 359, 367, 374, 376–8, 413, 455, 458, 670, 711, 746.
22. Alan K. Bowman, Peter Garnsey and Dominic Rathbone, eds., *The Cambridge Ancient History Volume 11: High Empire, A.D. 70–192* (vol. XI, Cambridge, 2000), pp. 72, 80, 149, 226, 252, 551, 622, 825.
23. Alan Bowman, Averil Cameron and Peter Garnsey, eds., *The Cambridge Ancient History Volume 12: Crisis of Empire, AD 193–337* (Cambridge, 2005), pp. 4, 77, 82, 133, 162–3, 175–7, 180, 197, 276, 283–4, 318, 324, 333, 336, 341, 347, 351, 356, 367, 371, 378–80, 390.
24. Averil Cameron and Peter Garnsey, *The Cambridge Ancient History Volume 13: The Late Empire, AD 337–425* (Oxford, 1998), pp. 66, 68, 73, 214.

25. Gary Forsythe, 'The Army and the Centuriate Organisation in Early Rome', in P. Erdkamp, ed., *A Companion to the Roman Army* (Oxford, 2007), pp. 28, 31.
26. Keppie, *The Making of the Roman Army*, p. 19, but see Note 54.
27. Pierre Cagniart, 'The Late Republican Army (146–30BC)', in P. Erdkamp, ed., *Companion to the Roman Army* (Oxford, 2007), pp. 87, 91.
28. Goldsworthy, *The Complete Roman Army*, p. 47.
29. Webster, *The Roman Imperial Army*, p. 19. This is also found in Webster's 1969 edition on p.37.
30. R. Alston, *Aspects of Roman History, AD 14–117* (London, 1998), p. 265.
31. K. Gilliver, 'The Augustan Reform and the Structure of the Imperial Army', in P. Erdkamp, ed., *A Companion to the Roman Army* (Oxford, 2007), Ch. 11.
32. J.H. Farnum, *The Positioning of the Roman Imperial Legions* (Oxford, 2005), p. 3.
33. K. Gilliver, 'The Augustan Reform and the Structure of the Imperial Army', in P. Erdkamp, ed., *A Companion to the Roman Army* (Oxford, 2007), pp. 183–5.
34. For example: Brizzi refers to *réforme* 24 times: Giovanni Brizzi, *Le guerrier de l'antiquité classique : de l'hoplite au légionnaire* (Monaco, 2004). Leonhard Burckhardt, *Militärgeschichte der Antike* (2008), p. 103. Le Bohec uses *réforme* thirty-four times: Yann Le Bohec, *Histoire des guerres romaines milieu du VIIIe siècle avant J.-C.-410 après J.-C.* (Paris, 2017).
35. For example, a Premier of NSW, Morris Iemma, describing his government's response to concerns about donations to political parties stated on 28 February 2008 in the Legislative Assembly in relation to Campaign Finance Laws: 'What we will do Mr Speaker is ensure the reforms work. That there is reform, not just change'. This shows the enduring sense of reform as something more than mere change: http://www.parliament.nsw.gov.au/prod/parlment/hanstrans.nsf/V3ByKey/LA20080228 (accessed 7 April 2008). Michael Evans writing in *Quadrant* referred to Prime Minister Howard's government 'pursuit of Army reform': M. Evans, 'Securing Australia's "Special Intersection"', *Quadrant* 446 (2008): p. 11.
36. There is a much-debated example of the use of this last term in the modern accounts of Augustus' constitutional arrangements. Syme presents Augustus as an artful manipulator of people, circumstances and images. In describing the constitutional arrangements of 28 and 27 BCE, Syme claims that the 'official language' was *res publica reddita* or *res publica restituta*: R. Syme, *The Roman Revolution* (London, 1939 (r.p.1960)), p. 323. These uses are suspect because they come from a disputed restoration of the Fasti Praenestini: E.A. Judge, 'Second Thoughts on Augustus', *Ancient Society*, 27 (1997), p. 52. Judge does not dispute the use of the term *res publica restituta* but he believes that it is not used in a constitutional sense. See also E. Gruen and K. Galinsky, 'Augustus and the Making of the Principate', *The Cambridge Companion to the Age of Augustus* (Cambridge, 2005), p. 34.
37. Ov. *Met.* 9.399, 11.254. G.B. Ladner, *The Idea of Reform, its Impact on Christian Thought and Action in the Age of the Fathers* (Cambridge, Mass., 1959), p. 39.
38. Dion Hal.. Sic. 4.81, metamorphosis of Actaeon.
39. Val. Max. 6.5.
40. Sen. *Ep.* 58.26.
41. G. Alföldy, 'The crisis of the third century as seen by contemporaries', *Greek, Roman, and Byzantine Studies*, 15 (1974), pp. 89–111.
42. Livy 8.8.3.

43. Fernando Echeverría Rey, 'Weapons, Technological Determinism and Ancient Warfare', in Garrett G. Fagan and Matthew Trundle, eds., *New Perspectives on Ancient Warfare* (Leiden, 2010), pp. 29–30., Xen. Eq. mag. 12.112; Amm. 24.6.8, 25.1.12–13; Pol. 6.22–23, 18.29–30.
44. Ladner, *The Idea of Reform, its Impact on Christian Thought and Action in the Age of the Fathers*, p. 16f. Echeverría Rey, 'Weapons, Technological Determinism and Ancient Warfare', pp. 27, 32.
45. Ladner, *The Idea of Reform, its Impact on Christian Thought and Action in the Age of the Fathers*, p. 15.
46. Tert. *DPal.* 2.7.
47. Alföldy, 'The crisis of the third century as seen by contemporaries', p. 110.
48. *Ibid.*, p. 3.
49. Ladner, *The Idea of Reform, its Impact on Christian Thought and Action in the Age of the Fathers*, p. 156f.
50. Joanna Innes, '"Reform" in English public life: the fortunes of a word', in Arthur Burns and Joanna Innes, eds., *Rethinking the Age of Reform* (Cambridge, 2003), p. 74. Ladner, *The Idea of Reform, its Impact on Christian Thought and Action in the Age of the Fathers*, pp. 41–5, 47–8, 133–283.
51. *Ibid.*, p. 26.
52. J. B. Bury, *The Idea of Progress: An Inquiry into Its Origin and Growth* (London, 1920), p. 21.
53. *Ibid.*, Ch. IV and R. Cawdrey, 'A Table Alphabeticall, conteyning and teaching the understanding of hard usuall English wordes', Ian Lancashire, revised from transcription by Raymond Siemens, http://leme.library.utoronto.ca/lexicons/record.cfm?id=276 (accessed 2 April 2008).
54. *Webster's Revised Unabridged Dictionary (1913 + 1828)*, http://machaut.uchicago.edu/websters (accessed 3 April 2008).
55. Ladner, *The Idea of Reform, its Impact on Christian Thought and Action in the Age of the Fathers*, p. 2.
56. *Ibid.*, p. 31.
57. Pascal Firges, Johna Lange, Thomas Maissen, Sebastian Meurer, Susan Richter, Gregor Stiebert, Lina Weber and Christine Zabel, 'Languages of reform in the European Enlightenment', in Susan Richter, Thomas Maissen and Manuela Albertone, eds., *Languages of Reform in the Eighteenth Century: When Europe Lost Its Fear of Change* (New York, 2020), pp. 1–6.
58. E. J. Hobsbawm, *Industry and Empire: from 1750 to the Present Day* (Harmondsworth, 1969), pp. 238–44.
59. M. Salvadori, *The Liberal Heresy* (London, 1977), p. 86; Innes, '"Reform" in English public life: the fortunes of a word', p. 89.
60. The reforms were a number of separate measures: A.V. Tucker, 'Army and Society in England 1870–1900: a Reassessment of the Cardwell Reforms', *The Journal of British Studies*, 2 (1962), p. 113. Lord Haldane introduced to the British Army further changes, which are also known as reforms, in the first decade of the Twentieth Century: Bryce Poe II, 'British Army Reforms 1902–1914', *Military Affairs*, 31 (1967), pp. 131–8.
61. Echeverría Rey, 'Weapons, Technological Determinism and Ancient Warfare', pp. 27–8. Nisbet argues that examples of progressive thinking can be found in Graeco-Roman

texts, but the result can best be described as cherry picking and special pleading: R.A. Nisbet, *History of the Idea of Progress* (1980).
62. Sallust, *Cat.* 23.4–24.1, cf. 21.3; Appian, *BC*, 2.2; Plut. *Vit. Cic.* 10–1.
63. Cic. *De Off.* 2.84.
64. Thuc. 1.1ff, 1.7.1.3. See Echeverría Rey, 'Weapons, Technological Determinism and Ancient Warfare', p. fn.16.
65. A. Momigliano, *Essays in Ancient and Modern Historiography* (Oxford, 1977), p. 194. Dodds believes that there is some evidence that the Ionians and the Athenians of the fifth century BC had some sense of progress, but this view did not mature into a new view for all Greeks: E.R. Dodds, *The Ancient Concept of Progress* (Oxford, 1973), p. 6f.
66. *Ibid.*, p. 18f.
67. Plato *Leg.* 678–9.
68. Plato *Prt.* 320d–321d.
69. Lucr. 2.1150 (*fracta est aetas*) and 5.392–431. Note also 3.945.
70. S. Mazzarino, *The End of the Ancient World* (London, 1966), p. 22.
71. Cic. *Rep.* 5.1.2.
72. Mazzarino, *The End of the Ancient World*, p. 27. Ovid does not show time as degenerative, notwithstanding Janus' quip, and Ovid presents time in his calendar as a constant re-enactment: Ov. *Met.* 1.24, 33.
73. Marcus Aurelius 11.1.2
74. Diod. 16.3.2; Frontin. *Str.* 3.*pr.*
75. J. E. Lendon, *Soldiers and Ghosts: A History of Battle in Classical Antiquity* (New Haven, 2005), pp. 11–13. Dodds, *The Ancient Concept of Progress*, pp. 24–5.
76. George Raudzens, 'War-Winning Weapons: The Measurement of Technological Determinism in Military History', *Journal of Military History*, 54 (1990), p. 405.
77. Organizationally, it can be argued that reforms are not changes but attempts at change: Nils Brunsson, *Reform as Routine: Organizational Change and Stability in the Modern World* (Oxford, 2009), p. 6.
78. Ross Cowan and Adam Hook, *Roman Battle Tactics, 109 BC–AD 313* (Oxford, 2007), pp. 57–8. See also the discussion in A. Goldsworthy, *Roman Warfare* (London, 2000), pp. 167–9. Goldworthy sees some variations across time in Roman infantry tactics but he does not see them as substantial.
79. Thuc. 1.23.
80. See the comments on the debate about Roman strategy: Everett L Wheeler, 'The Army and the Limes in the East', in P. Erdkamp, ed., *A Companion to the Roman Army* (Oxford, 2007), pp. 237–8.
81. P. Garnsey and R. Saller, *The Roman Empire: Economy, Society, Culture* (London, 1987), Ch. 2.
82. N. J. E Austin and N. B Rankov, *Exploratio: Military and Political Intelligence in the Roman World from the Second Punic War to the Battle of Adrianople* (London, 1995), pp. 135–6.
83. Theodor Mommsen and Joachim Marquardt, *Römisches Staatsrecht* (vol. 1, Graz, 1952), I, p. 116. Cic. *Phil.* 5.45.
84. R.G.M. Nisbet, 'Aeneas Imperator: Roman Generalship in an Epic Context', *Oxford Readings in Virgil's Aeneid*, 18 (1990), pp. 378–89. J. Sarkissian, 'The Idea of Imperium in Aeneid 1, 50–296', *The Augustan Age*, 4 (1985), pp. 51–6.

85. Mommsen and Marquardt, *Römisches Staatsrecht*, pp. 119–30.
86. Catherine. M. Gilliver, 'The Roman Army and the Morality of War', in A.B. Lloyd, ed., *Battle in Antiquity* (London, 1996), p. 220.
87. Cic. *Leg. Man.* 10.28.
88. Cic. *Leg.Man.* 16.49.
89. Cicero in the same speech also refers to *consilium* in the sense of sound decision-making or planning: *Ibid.* 20.60. Pompey acts *divino consilio*: Cic. Leg. Man. 10. Later, Cicero extols Pompey: *quantum consilio, quantum dicendi gravitate et copia*: Cic. Leg. Man. 42.
90. Livy 44.22.
91. Pliny *Ep.* 8.14.4–5, 7. See also Tac. *Dial.* 34.1 and Sen. *Ep.* 1.6.5.
92. Cic. *Font.* 42–3.
93. Cic. *Balb.* 47.
94. J.B. Campbell, 'Teach yourself how to be a general', *Journal of Roman Studies*, 77 (1987), pp. 13–28.
95. Cincius the Antiquarian is referred to as a military writer in Aul. Gel. *NA* 16.4.
96. Ael. *Tact.* 33f.
97. Veg. *Mil.* 1.8.
98. Campbell is surprised by the 'limited concept of the qualities and skills required of a military commander': Campbell, 'Teach yourself how to be a general', *Journal of Roman Studies, Ibid.*
99. Onasander Strategikos P.1. For Veranius see R. Syme, 'The Origin of the Veranii', *The Classical Quarterly*, New Series, 7 (1957), pp. 123–25.
100. Onasander 25.
101. ILS 2487 and Arr. *Tactica* 32.3.
102. Arr. *Tactica* 32.3.
103. Cass. Dio 69.9.
104. Veg. *Mil.* praefatio 15.
105. On stratagems, see Everett L. Wheeler, *Stratagem and the Vocabulary of Military Trickery* (New York, 1988).
106. Frontin. *Str.* 1.
107. For example: Julius Africanus, Κεστοί 38, 45 (use of fire); 46 (burning gates); 50 (smuggling in weapons); 51–52 (sending secret messages).
108. Polyaenus *Strat.* 16.1–2.
109. Caes. *BCiv.* 1.27.
110. Caes. *BGal.* 1.46.
111. Caes. *BCiv.* 3.75.
112. Caes. *BAfr.* 77.
113. Caes. *BCiv.* 1.44.
114. Caes. *BAfr.* 20.

Chapter 4

1. Much of the content of this chapter has been published in a different form in A.A. McArthur, 'Should Roman Soldiers be called "Professional" prior to Augustus?', *Journal of Military History*, 85:1 (2021), pp. 9–26.
2. Keaveney does not refer to professionalism: Keaveney, *The Army in the Roman Revolution*. Lendon focuses more on *disclipina* and *virtus*, although he acknowledges

professionalism in the imperial period: Lendon, *Soldiers and Ghosts: A History of Battle in Classical Antiquity*, p. 232. Phang notes the difficulties of using professionalism in relation to Rome's armies: S.E. Phang, *Military Service. Ideologies of Discipline in the Late Republic and Early Principate* (Cambridge, 2008), pp. 3, 37, 73, 82f. Rosenstein uses the term but notes the limitations: N. Rosenstein, 'Military Command, Political Power, and the Republican Elite', in P. Erdkamp, ed., *A Companion to the Roman Army* (Oxford, 2007), pp. 132, 144, 167f. Rich does not believe that Marius 'gave Rome a professional army': J. Rich, 'The supposed Roman manpower shortage of the later second century BC', *Historia*, 32 (1983), p. 232. See also: Colin Adams, 'War and Society', in Philip Sabin, Hans Van Wees and Michael Whitby, eds., *The Cambridge History of Greek and Roman Warfare: Volume II Rome from the Late Republic to the Late Empire* (Cambridge, 2007), pp. 198, 209.
3. K. Strobel, 'Strategy and Army Structure between Septimius Severus and Constantine the Great', in P. Erdkamp, ed., *A Companion to the Roman Army* (Oxford, 2007), pp. 268, 273. Adams, 'War and Society', p. 213.
4. D.J.B. Trim, ed., *The Chivalric Ethos and the Development of Military Professionalism* (Leiden, 2003), p. 4.
5. Sarah Williams and Alan Apperley, 'Public Relations and Discourses of Professionalism', in A. Rogojinaru and S. Wolstenholme, eds., *Current Trends in International Public Relations* (Bucharest, 2009), p. 4. and Peter. Feaver, *Armed Servants: Agency, Oversight, and Civil-Military Relations* (London, 2003), p. 89.
6. Julia Evetts, 'Explaining the Construction of Professionalism in the Military: History, Concepts and Theories', *Ophrys: revue française de sociologie*, 44 (2003/4), pp. 760–61.
7. *Ibid.*, p. 769.
8. *Ibid.*
9. Thomas-Durell Young, 'Military Professionalism in a Democracy', in Thomas C. Bruneau and D. Tollefson Scott, eds., *Who Guards the Guardians and How* (Austin, 2006), p. 2. Trim, *The Chivalric Ethos and the Development of Military Professionalism*, pp. 5–7. Col USA (Ret) Richard Swain, *The Obligations of Military Professionalism: Service Unsullied by Partisanship* (Washington, 2010), p. 3. Sam C. Sarkesian and Connor, Robert E. Jr, *The US Military Profession into the Twenty-First Century: War, Peace and Politics* (Oxford, 2006), p. 27. Samuel P. Huntington, *The Soldier and the State: The Theory and Politics of Civil-Military Relations.* (Cambridge, 1957), pp. 19f, 30–1. David R. Gray, 'New Age Military Progressives: U.S. Army Officer Professionalism in the Information Age', (U.S. Army War College, Carlisle Barracks, Pennsylvania, 2001), pp. 1, 5.
10. Jeffery A. Bradford, 'Proconsuls and CinCs from the Roman Republic to the Republic of the United States of America: Lessons for the Pax Americana', (Monograph, School of Advanced Military Studies United States Army Command and General Staff College Fort Leavenworth, Kansas, 2001), pp. 27–8.
11. M. van Crefeld, *The Art of War: War and Military Thought* (London, 2000), p. 131f.
12. Evetts, 'Explaining the Construction of Professionalism in the Military: History, Concepts and Theories', p. 760. Young, 'Military Professionalism in a Democracy', p. 19. Bradford, *Proconsuls and CinCs from the Roman Republic to the Republic of the United States of America: Lessons for the Pax Americana*, pp. 27–8.
13. Edward Gibbon and Georges Alfred Bonnard, *Memoirs of my Life* (London, 1966), p. 117.

14. Young, 'Military Professionalism in a Democracy', pp. 3, 19.
15. *Ibid.*, pp. 5, 7, 18–19, 23. Swain, *The Obligations of Military Professionalism: Service Unsullied by Partisanship*, p. 4. Gray, *New Age Military Progressives: U.S. Army Officer Professionalism in the Information Age*, p. 17. S. E. Finer, *The Man on Horseback: The Role of the Military in Politics* (New Brunswick, 2002), p. 28.
16. A. Cobban, *A History of Modern France* (Hammondsworth, 1965), vol III p. 48f.
17. Peter Simkins, *Kitchener's Army: The Raising of the New Armies, 1914–16* (Manchester, 1988), pp. 21–3.
18. Feaver, *Armed Servants: Agency, Oversight, and Civil-Military Relations*, p. 78.
19. Sarkesian and Connor, Robert E. Jr, *The US Military Profession into the Twenty-First Century: War, Peace and Politics*, p. 25.
20. John A. Lynn, *Battle: A History of Combat and Culture* (New York, 2003), p. 470.
21. Georg G. Iggers, Q. Edward Wang and Supriya Mukherjee, *A Global History of Modern Historiography* (London, 2008), p. 22f. John Childs, *Armies and Warfare in Europe, 1648–1789* (Manchester, 1982), p. 91f. Yvon Garlan, *War in the Ancient World: A Social History* (London, 1975), p. 19.
22. J. Ravenel, *Oeuvres complètes de Montesquieu* (Paris, 1834).
23. Edward Gibbon, *The History of the Decline and Fall of the Roman Empire, Volume the Second* (vol. II, London, 1781), II, pp. 48–9, 214. Edward Gibbon, *The History of the Decline and Fall of the Roman Empire, Volume the Third* (vol. III, London, 1781), III, pp. 115, 106, 208, 302, 390, 453. Edward Gibbon, *The History of the Decline and Fall of the Roman Empire, Volume the Fourth* (vol. IV, London, 1788), IV, pp. 43, 352, 495. Edward Gibbon, *The History of the Decline and Fall of the Roman Empire, Volume the Fifth* (vol. V, London, 1788), V, p. 57. Edward Gibbon, *The History of the Decline and Fall of the Roman Empire, Volume the Sixth* (vol. VI, London, 1788), VI, pp. 65, 215, 451.
24. Adam Ferguson, *The History of the Progress and Termination of the Roman Republic.* (vol. III, London, 1783), III, pp. 452, 515.
25. Hooke, *The Roman History from the Building of Rome to the Ruin of the Commonwealth*, Vols II and III (Dublin, 1759)
26. Goldsmith, *Goldsmith's Roman History: Abridged by Himself, for the Use of Schools*.
27. B.G. Niebuhr, *Römische Geschichte, erster Theil* (Berlin, 1828). B.G. Niebuhr, *Römische Geschichte, zweiter Theil* (Berlin, 1812). B.G. Niebuhr, *Römische Geschichte, dritter Theil* (Berlin, 1832). B.G. Niebuhr, *Römische Geschichte, vierter Theil* (Jena, 1844).
28. Michelet, Jules, *Histoire romaine, première partie: République* (vol. 1, Paris, 1834), I, p. 170. Jules Michelet, *Histoire romaine. 1ère partie: République* (vol. 2, Paris, 1843), II, p. 53.
29. M. Todière, *Sommaire d'un cours complet d'histoire romaine* (Tours, 1846).
30. Arnold, *The History of Rome*, pp. 15, 306.
31. T. Arnold, *History of Rome, Volume II* (vol. II, London, 1848), II p. 558.
32. Theodor Mommsen, *Römische Geschichte, Erster Band bis zur Schlacht von Pydna* (vol. I, Berlin, 1856).
33. Theodor Mommsen, *Römische Geschichte, zweiter Band von der Schlacht bei Pydna bis auf Sullas Tod* (vol. II, Berlin, 1857), II, pp. 168, 217, 360.
34. *Ibid.*, p. 217.
35. Mommsen, Theodor, *Römische Geschichte: die Provinzen von Caesar bis Diocletian.* (Berlin, 1885).

36. Theodor Mommsen, *The History of Rome, Volume IV* (New York, 1871), IV, pp. 268, 356, 579, 597, 606, 608, 634, 671, 709, 728, 730, 732–3.
37. T.A. Dodge, *Caesar: A History of the Art of War Among the Romans Down to the End of the Roman Empire* (New York, 1892), pp. 1, 352.
38. T. Mommsen, *The History of Rome, Volume III* (London, 1863).
39. Leighton, *A History of Rome*, p. 365. Leighton later states that: "Under the empire the army became a permanent organisation", p. 379.
40. Livy 5.7.5. Myers, *A History of Rome*, pp. 41–42.
41. Delbrück, *Geschichte der Kriegskunst im Rahmen der politischen Geschichte: das Altertum*, p. 433.
42. Frank, *A History of Rome*, p. 91.
43. Kubitschek, 'Legio. Republikanische Zeit', col. 1204.
44. Kromayer and Veith, *Heerwesen und Kriegsführung der Griechen und Römer*, p. 295.
45. *Ibid.*, pp. 219, 314, 316, 485.
46. H.M.D. Parker, *The Roman Legions* (Oxford, 1928), p. 25.
47. H. Last, 'Wars of the Age of Marius', in S.A. Cook, F.E. Adcock and M.P. Charlesworth, eds., *The Cambridge Ancient History Volume IX: the Roman Republic* (London, 1932), p. Preface v.
48. *Ibid.*, p. 36.
49. H. Last, 'Gaius Gracchus', in S.A. Cook, F.E. Adcock and M.P. Charlesworth, eds., *The Cambridge Ancient History Volume IX: the Roman Republic* (London, 1932), p. 48.
50. Last, 'Wars of the Age of Marius', p. 136.
51. *Ibid.*, p. 137.
52. Cary, *A History of Rome down to the Reign of Constantine*, p. 308.
53. H. H. Scullard, *A History of the Roman World from 753 to 146 B.C* (London, 1951), p. 100.
54. Syme, *The Roman Revolution*, p. 395.
55. F. E. Adcock, *The Roman Art of War under the Republic* (Cambridge, Mass, 1940), pp. 17–19. Dawson holds the same view of the importance of centurions: Doyne Dawson, *The Origins of Western Warfare: Militarism and Morality in the Ancient World* (Boulder, Colo., 1996), p. 112.
56. E. Gabba, *Republican Rome, the Army and the Allies* (Oxford, 1976), p. Ch. 2.
57. . '…l'année 107 vu les vielles armées du type censitaire céder la place à une armée nouvelle à recrutement prolétarien, ceci entraînant le replacement, par les troupes professionnelle et permanentes, des levées de l'ancien système, licenciées après chaque campagne', i.e. the year 107 saw the old armies of the censitaire type give way to a new army with proletarian recruitment, this leading to the replacement, by the professional and permanent troops, of the levies of the old system, dismissed after each campaign , J. Harmand, *L'Armée et le soldat à Rome de 107 à 50 avant notre ère* (Paris, 1967), pp. 387–396.
58. P.A. Brunt, *Italian Manpower* (Oxford, 1971), p. 310.
59. Smith, *Service in the Post-Marian Roman Army*, pp. 3, 5, 35.
60. *Ibid.*, p. 8.
61. Brunt, *Italian Manpower*, p. 396. For Ligustinus, see Livy 42, 32–4.
62. Brunt, *Italian Manpower*, p. 399.
63. Garlan, *War in the Ancient World: a Social History*, pp. 103f, 106.

64. Keppie, *The Making of the Roman Army*, pp. 51, 53, 55. Lawrence Keppie, 'The Army and the Navy', in Alan K. Bowmans, Edward Champlin and Andrew Lintott, eds., *The Cambridge Ancient History: Volume X The Augustan Empire, 43 B.C.-A.D. 69* (Cambridge, 1996), p. 371.
65. Keppie, *The Making of the Roman Army*, p. 59.
66. Campbell, 'Teach yourself how to be a general', p. 23.
67. Webster, *The Roman Imperial Army*, p. 16.
68. John Patterson, 'Military organisation and social change in the later Roman Republic', in John Rich and Graham Shipley, eds., *War and Society in the Roman World* (London, 1993), p. 99.
69. *Ibid.*, p. 104.
70. Dawson, *The Origins of Western Warfare: Militarism and Morality in the Ancient World*, p. 118.
71. Webster, *The Roman Imperial Army*, p. 27.
72. *Ibid.*, p. 142.
73. Adrian Keith. Goldsworthy, *Roman Warfare*, ed. John Keegan (London, 2000), pp. 18, 92f.
74. *Ibid.*, p. 49.
75. Goldsworthy, *The Roman Army at War: 100 BC-AD 200*, p. 34.
76. Goldsworthy, *Roman Warfare*, p. 78.
77. M. Feugère, *Les armes des romains de la république à l'antiquité tardive* (Paris, 2002), p. 45.
78. Erik Hildinger, *Swords Against the Senate: The Rise of the Roman Army and the Fall of the Republic* (Cambridge Mass., 2002), pp. 99, 117–18, 144.
79. Lynn, *Battle: A History of Combat and Culture*, p. 46.
80. Brizzi, *Le guerrier de l'antiquité classique : de l'hoplite au légionnaire*, p. 158.
81. *Ibid.*, p. 179. Nathan Stewart. Rosenstein, *Rome at War: Farms, Families, and Death in the Middle Republic* (Chapel Hill, N.C., 2004), p. 83; Erdkamp, *A Companion to the Roman Army*, p. 2.
82. L. Rawlings, 'Army and Battle During the Conquest of Italy (350–264 BC)', in P. Erdkamp, ed., *A Companion to the Roman Army* (Oxford, 2007), p. 58. D. Hoyos, 'The Age of Overseas Expansion (264–146 B.C.)', in P. Erdkamp, ed., *A Companion to the Roman Army* (Oxford, 2007), p. 64.
83. *Ibid.*, p. 76.
84. Cagniart, 'The Late Republican Army (146–30BC)', pp. 80, 82–3, 91, 93.
85. *Ibid.*, p. 84.
86. Lukas de Blois, 'Army and General in the Late Roman Republic', in P. Erdkamp, ed., *A Companion to the Roman Army* (Oxford, 2007), p. 166.
87. *Ibid.*, p. 167.
88. *Ibid.*, pp. 168–9, 174.
89. Richard Alston, 'The Military and Politics', in Philip Sabin, Hans Van Wees and Michael Whitby, eds., *The Cambridge History of Greek and Roman Warfare: Volume II Rome from the late Republic to the late Empire* (Cambridge, 2007), p. 180.
90. Gilliver, 'The Augustan Reform and the Structure of the Imperial Army', p. 183f.
91. Pierre Cosme, *L'armée romaine : VIIIe s. av. J.-C.-Ve s ap. J.-C* (Paris, 2012), p. 56.
92. Ibid., p. 79.
93. Pat Southern and Karen R. Dixon, *The Late Roman Army* (London, 1996), p. 96.
94. Southern, *The Roman Army: A Social and Institutional History*, pp. 96, 125, 128.
95. *Ibid.*, p. 201.

96. Kathryn H. Milne, 'The Republican Soldier: Historiographical Representations and Human Realities', (Ph.D. thesis, University of Pennsylvania, 2009), p. 150.
97. *Ibid.*, p. 164.
98. Jonathan P. Roth, *Roman Warfare* (Cambridge, 2009), pp. 91, 134.
99. Christopher Anthony Matthew, *On the Wings of Eagles: The Reforms of Gaius Marius and the Creation of Rome's First Professional Soldiers* (Newcastle upon Tyne, 2010).
100. See Erich. S. Gruen, *The Last Generation of the Roman Republic* (Berkeley, 1995), p. xvii.
101. Livy 42.32.
102. Livy 42.34.
103. Pol. 6.20f.
104. M.C.J. Miller, 'The Professionalization of the Roman Army in the Second Century B.C.', (Ph. D. thesis, Chicago, 1984), p. 3.
105. Duffy, *The Military Experience in the Age of Reason*, pp. 100–101.
106. D. Woodward, *The Armies of the World 1854–1914* (London, 1978), p. 108.
107. Rosenstein, 'Military Command, Political Power, and the Republican Elite', p. 132.
108. Adcock, *The Roman Art of War under the Republic*, p. 17.
109. Cadiou, *L'armée imaginaire: les soldats prolétaires dans les légions Romaines au dernier siécle de la République*.
110. Gruen, *The Last Generation of the Roman Republic*, p. xvii.
111. Cadiou, *L'armée imaginaire: les soldats prolétaires dans les légions Romaines au dernier siécle de la République*, p. 46.
112. Keaveney, *The Army in the Roman Revolution*.
113. Werner Eck, *Augustus und seine Zeit*, ed. Eck Werner (C. H. Beck Wissen, vol. 2084, München, 2014).
114. S.P. Oakley, 'Single Combat in the Roman Republic', *The Classical Quarterly*, New Series 35 (1985), p. 407.
115. Lendon, *Soldiers and Ghosts: A History of Battle in Classical Antiquity*, p. 169.
116. Livy 5.7.5.
117. Keaveney, *The Army in the Roman Revolution*, p. 64.
118. Lendon, *Soldiers and Ghosts: A History of Battle in Classical Antiquity*, p. 231.
119. *Ibid.*, p. 236.
120. *Ibid.*, p. 255.
121. *Ibid.*, Ch. 13.

Chapter 5

1. See also Livy. Praef 6–7, 8.4.
2. Guy Jolyon Bradley, *Early Rome to 290 BC: The Beginnings of the City and the Rise of the Republic* (Edinburgh, 2020), p. 8f.
3. *Ibid.*, p. 15.
4. *Ibid.*, p. 6.
5. K.A. Raaflaub, 'Between Myth and History: Rome's Rise from Village to Empire (the Eight Century to 264)', in Nathan Rosenstein, and Robert Morstein-Marx, eds., *A Companion to the Roman Republic* (Malden, 2006), p. 128.
6. Bradley, *Early Rome to 290 BC: The Beginnings of the City and the Rise of the Republic*, p. 17.
7. Cic. Leg. 1.6. Livy 8.40.3–5

8. G. Forsythe, *A Critical History of Early Rome* (Berkeley, 2005), p. 63.
9. Goldsworthy, *Roman Warfare*, p. 26.
10. E.g., Dominique. Briquel, *Mythe et révolution: a fabrication d'un récit : la naissance de la république à Rome* (Bruxelles, 2007), pp. 328, 336, Forsythe, 'The Army and the Centuriate Organisation in Early Rome', p. 7, Raaflaub, 'Between Myth and History: Rome's Rise from Village to Empire (the Eight Century to 264)', p. 133, Michael M. Sage, *The Army of the Roman Republic: From the Regal Period to the Army of Julius Caesar* (Barnsley, South Yorkshire, 2018), Ch. 1.
11. Timothy J. Cornell, *The Beginnings of Rome: Italy and Rome From the Bronze Age to the Punic Wars (C 1000–264 BC)* (London, 1995), p. xv, 10. 15. E. Rawson, 'The Literary Sources for the Pre-Marian Army', *Papers of the British School at Rome*, 39 (1971), p. 13. Drogula, *Commanders and Command in the Roman Republic and Early Empire*, p. 11f.
12. M.T. Burns, 'The Homogenisation of Military Equipment under the Roman Republic', *www.digressus.org*, Digressus Supplement 1 'Romanisation?' (2003), p. 60.
13. Bradley, *Early Rome to 290 BC: The Beginnings of the City and the Rise of the Republic*, p. 26.
14. *Ibid.*, p. 101.
15. The best general survey on the early sources on the Roman army is E. Rawson, 'The Literary Sources for the Pre-Marian Army,' *Papers of the British School at Rome*, 39 (1971): pp. 13–31.
16. Livy 1.43. Dion. Hal. *Ant. Rom.* 4.16.
17. Sekunda et al., *Early Roman Armies*, 6. Additional sources can be found in Cic. *De Rep.* 2, 39–40; POxy. 2088; Plin. *Nat. Hist.* 18, 13; Eutrop. 1,7.
18. Livy 1.42.
19. *Ibid.* 1.43.
20. Dion. Hal. *Ant. Rom.* 5.75.4ff.
21. Goldsworthy, *Roman Warfare*, p. 35. Sage, *The Army of the Roman Republic: From the Regal Period to the Army of Julius Caesar*, Ch. 2.
22. Cornell, *Early Rome to 290 BC: The Beginnings of the City and the Rise of the Republic (C 1000–264 BC)*, p. 180.
23. Bradley, *Early Rome to 290 BC: The Beginnings of the City and the Rise of the Republic*, p. 106. Cornell, *The Beginnings of Rome: Italy and Rome From the Bronze Age to the Punic Wars (C 1000–264 BC)*, p. 181.
24. Varro *Ling.* 6.93.
25. Cass. Dio 37.28.
26. Cornell, *The Beginnings of Rome: Italy and Rome from the Bronze Age to the Punic Wars (C 1000–264 BC*, p. 182.
27. H. Last, 'The Servian Reforms', *Journal of Roman Studies*, 35 (1945): pp. 34–5, 44.
28. Livy 1.42.
29. Dion. Hal. *Ant. Rom.* 4.15.1, 4.14.5.
30. Livy 6.1.2–3. See also J. Briscoe, 'The First Decade', in *Livy*, ed. T.A. Dorey (London: Routledge and Keegan Paul, 1971), pp. 4–5.
31. Last, 'The Servian Reforms', p. 38.
32. Marianna Scapini, 'Literary Archetypes for the Regal Period', in Bernard Mineo, ed., *A Companion to Livy (*Chichester, 2015), p. 279.
33. Forsythe, *A Critical History of Early Rome*, p. 114.

34. Bradley, *Early Rome to 290 BC: The Beginnings of the City and the Rise of the Republic*, p. 193.
35. Forsythe, 'The Army and the Centuriate Organisation in Early Rome', p. 17. Le Bohec, *Histoire des guerres romaines milieu du VIIIe siècle avant J.-C.-410 après J.-C*, p. 62.
36. Dion. Hal. *Ant. Rom.* 6.16.
37. Parker, H.M.D. Parker, *The Roman Legions* (London, 1958), p. 11. Keppie, *The Making of the Roman Army*, p. 19. Delbrück, *Warfare in Antiquity: History of the Art of War*, p. 272. Webster, *The Roman Imperial Army*, p. 3f.
38. R.H. Storch, 'The Archaic Greek Phalanx 750–650 B.C.', *The Ancient History Bulletin*, 12 (1998), p. 7.
39. John Rich and Graham Shipley, eds., *War and Society in the Roman World* (2002), p. 17.
40. John. Rich, 'Warfare and the army in early Rome', in Paul Erdkamp, Anthony Birley, William Broadhead, Lukas De Blois, Pierre Cagniart, Hugh Elton, Gary Forsythe, Luuk De Ligt and Clifford Ando, eds., *A Companion to the Roman Army* (Oxford, 2007), p. 17.
41. See the discussion in N. Sekunda, S. Northwood and R. Hook, *Early Roman Armies* (London, 1995), p. 34f.
42. Michael J. Taylor, 'Etruscan identity and service in the Roman army: 300–100 B.C.E.', *American Journal of Archaeology: The Journal of the Archaeological Institute of America*, 121 (2017), pp. 281, 287–290.
43. F.H. Massa-Pairault, 'Notes sur le problème du citoyen en armes: cité romaine et cité étrusque', in A. Rouveret A.-M. Adam, ed., *Guerre et sociétés en Italie aux Ve et IVe Siècles avant J-C* (Paris, 1986), pp. 29–50.
44. A. Rouveret A.-M. Adam, ed., *Guerre et sociétés en Italie aux Ve et IVe Siècles avant J-C* (Paris, 1986), p. 9.
45. Bradley, *Early Rome to 290 BC: The Beginnings of the City and the Rise of the Republic*, pp. 128–30.
46. *Ibid.*, p. 18.
47. Paul Erdkamp, 'Army and Society', in Nathan Rosenstein, and Robert Morstein-Marx, eds., *A Companion to the Roman Republic* (Malden, 2006), p. 282. Forsythe, *A Critical History of Early Rome*, p. 113.
48. Bradley, *Early Rome to 290 BC: The Beginnings of the City and the Rise of the Republic*, p. 284f.
49. Suet. *Tib.* 1; Livy 2.16–29; Dion. Hal. *Ant. Rom.* 5.40, 4; 23.24.2.
50. Livy 3.15. Bradley, *Early Rome to 290 BC: The Beginnings of the City and the Rise of the Republic*, p. 284.
51. *Ibid.*, p. 286f.
52. Goldsworthy, *Roman Warfare*, p. 44.
53. Oakley, 'Single Combat in the Roman Republic', pp. 106–107. Rich and Shipley, *War and Society in the Roman World*, p. 15.
54. Livy 5.37–48, Diod. 14.115–6, Plut. *Vit. Cam.* 22.1.
55. Livy 6.32.1.
56. Forsythe, *A Critical History of Early Rome*, p. 107.
57. Bradley, Bradley, *Early Rome to 290 BC: The Beginnings of the City and the Rise of the Republic*, p. 300.

58. *Ibid.*, p. 180.
59. A-M. Adam, 'Emprunts et échanges de certains types d'armement entre l'Italie et le monde non-méditerranéen aux Ve et IVe siècles avant J-C', in A. Rouveret and A.-M. Adam, ed., *Guerre et sociétés en Italie aux Ve et IVe Siècles avant J-C* (Paris, 1986), pp. 19–28; *Ibid.*, pp. 26–7.
60. Livy 4.59–60.
61. Massa-Pairault, 'Notes sur le problème du citoyen en armes: cité romaine et cité étrusque', pp. 40–43.
62. Michael H. Crawford, *Coinage and Money Under the Roman Republic: Italy and the Mediterranean Economy* (London, 1985), p. 22. James Tan, 'The *dilectus-tributum* system and the settlement of fourth century Italy.', in Jeremy Armstrong and Michael P. Fronda, eds., *Romans at War: Soldiers, Citizens, and Society in the Roman Republic* (London, 2020), p. 53.
63. *Ibid.*, p. 54.
64. Bradley, *Early Rome to 290 BC: The Beginnings of the City and the Rise of the Republic*, p. 325.
65. Kromayer and Veith, *Heerwesen und Kriegsführung der Griechen und Römer*, p. 262. Parker describes it as 'the next stage in the development of the army': Parker, *The Roman Legions*, p. 11. The 1998 edition of Webster's work on the Imperial Army accepts Camillus' changes as 'innovations' as in 1969 edition: Webster, *The Roman Imperial Army*, p. 6.
66. P. Couissin, *Les armes romaines* (Paris, 1926), p. 177. Parker, *The Roman Legions*, p. 11.
67. Parker, *The Roman Legions*, p. 11.
68. Miller claims that it was an ancient habit to attach a number of changes to a single reformer and that modern authors constructed both the reforms and the reformer. Miller also claims that Camillus was not responsible for the manipular reforms because he believes that the Roman disaster at the Caudine Forks was due to the continued retention of hoplite organization: M.C.J. Miller, 'The Principes and the so-called Camillan Reforms', *The Ancient World*, 23 (1992), pp. 68, 70. This position is difficult to accept given that the Romans appear to have surrendered without fighting after being trapped.
69. Keppie, *The Making of the Roman Army*, p. 19.
70. Cornell, *The Beginnings of Rome: Italy and Rome From the Bronze Age to the Punic Wars (C 1000–264 BC)*, p. 188.
71. Dion. Hal. *Ant. Rom.* 14.9.1–2.
72. Dion. Hal. *Ant. Rom.* 14.9.3.
73. Plut. *Vit. Cam.* 40.
74. Webster, *The Roman Imperial Army*, p. 24.
75. Pol. 2.33.4.
76. Miller, 'The Principes and the so-called Camillan Reforms', p. 68.
77. Forsythe, *A Critical History of Early Rome*, pp. 304–5.
78. Rawlings, 'Army and Battle During the Conquest of Italy (350–264 BC)', p. 54.
79. Juliusz Tomczak, 'Roman military equipment in the 4th century BC: pilum, scutum and the introduction of manipular tactics', *Acta Universitatis Lodziensis Folia Archaeologica*, 29 (2012), p. 38.

80. D. Briquel, 'La tradition sur l'emprunt d'armes Samite par Rome', in A. Rouveret and A.-M. Adam, ed., *Guerre et sociétés en Italie aux Ve et IVe Siècles avant J-C* (Paris, 1986), pp. 84–5.
81. Rich and Shipley, *War and Society in the Roman World*, p. 18. Massa-Pairault, 'Notes sur le problème du citoyen en armes: cité romaine et cité étrusque', p. 44.
82. *Ibid.*, pp. 46–7, quoting Livy 9.40 and Pol. 2.19.7–20.
83. Rawlings, 'Army and Battle During the Conquest of Italy (350–264 BC)', p. 55.
84. Livy 22.38.2–5.
85. Rawlings, 'Army and Battle During the Conquest of Italy (350–264 BC)', p. 57.
86. Parker, *The Roman Legions*, p. 12.
87. Livy 8.8; Burns, 'The Homogenisation of Military Equipment under the Roman Republic', p. 65.
88. H. von Arnim, 'Ineditum Vaticanum', *Hermes*, 27 (1892), pp. 118–130. See also Diod. 23.2.1; Athenaeus 6.273f; Sall. *Cat.* 51.37f.
89. Massa-Pairault, 'Notes sur le problème du citoyen en armes: cité romaine et cité étrusque', p. 58.; Dio. Hal. 23.2.1; Sall. *Cat.* 51. In support of Livy, see C.M. Gilliver, *The Roman Art of War* (Stroud, 1999), p. 15.
90. Briquel, 'La tradition sur l'emprunt d'armes Samite par Rome', p. 67.
91. Burns, 'The Homogenisation of Military Equipment under the Roman Republic', pp. 77–9. Feugère, *Les armes des romains de la république à l'antiquité tardive*, p. 87–88.
92. *Ibid.*, p. 89.
93. Cornell, *The Beginnings of Rome: Italy and Rome From the Bronze Age to the Punic Wars (C 1000–264 BC)*, p. 186.
94. Pol. 15.15.7.
95. Feugère, *Les armes des romains de la république à l'antiquité tardive*, p. 94.
96. *Ibid.*, p. 100. R. Cowan, 'Etruscan and Gallic Pila: origins of the heavy javelin', *Ancient Warfare*, XII-6 (2019), pp. 18–21.
97. Festus 224L s.v. *pulumnae poplae*. Forsythe, *A Critical History of Early Rome*, p. 181.
98. Olivier de Cazanove, 'Pratiques et rites de la guerre en Italie, entre Romains et Samnites: le passage sous le joug, la légion de lin samnite', in J-Chr. Couvenhes, S. Crouzet and S. Péré-Noguès, eds., *Pratiques et identités culturelles des armées hellénistiques du monde méditerranéen: Hellenistic Warfare 3* (Bordeaux, 2011), p. 369.
99. Rawlings, 'Army and Battle During the Conquest of Italy (350–264 BC)', p. 49.
100. Drogula, *Commanders and Command in the Roman Republic and Early Empire*, p. 31.
101. Forsythe, *A Critical History of Early Rome*, p. 211.
102. Drogula, *Commanders and Command in the Roman Republic and Early Empire*, p. 68.
103. Pol. 6.37.
104. Drogula, *Commanders and Command in the Roman Republic and Early Empire*, pp. 69, 80.
105. Val. Max. 1.4.3.
106. Drogula, *Commanders and Command in the Roman Republic and Early Empire*, pp. 125, 131, 139.
107. Arthur M. Eckstein, *Senate and General: Individual Decision Making and Roman Foreign Relations 264–194 B.C.* (London, 1987), p. 73.
108. Jeremiah McCall, 'The manipular army system and command decisions in the second century', in Jeremy Armstrong and Michael P. Fronda, eds., *Romans at War: Soldiers, Citizens, and Society in the Roman Republic* (London, 2020), p. 231.

109. Patrick Kent, 'Reconsidering «socii» in Roman armies before the Punic Wars', in Saskia T. Roselaar, ed., *Processes of integration and identity formation in the Roman Republic* (Leiden, 2012), pp. 77–81.
110. Jeremy Armstrong, 'Organised chaos: *Manipuli, socii,* and the Roman army c.300', in Jeremy Armstrong and Michael P. Fronda, eds., *Romans at War: Soldiers, Citizens, and Society in the Roman Republic* (London, 2020), pp. 94, 97–8.
111. Tan, 'The *dilectus-tributum* system and the settlement of fourth century Italy.', pp. 57–8.
112. *Ibid.*, p. 67.
113. *Ibid.*, pp. 59–60.
114. Rawlings, 'Army and Battle During the Conquest of Italy (350–264 BC)', p. 52.
115. Goldsworthy, *Roman Warfare*, p. 38. Kent, 'Reconsidering «socii» in Roman armies before the Punic Wars', p. 79–81. Rawlings, 'Army and Battle During the Conquest of Italy (350–264 BC)', p. 48.
116. Forsythe, *A Critical History of Early Rome*, p. 286.
117. Guy Bradley, *Early Rome to 290 BC: The Beginnings of the City and the Rise of the Republic*, p. 323.
118. Pol. 6.39.
119. Pol. 10.16. Nathan S. Rosenstein, 'Integration and Armies in the Middle Republic', in Saskia T. Roselaar, ed., *Processes of Integration and Identity Formation in the Roman Republic* (Boston (Mass.), 2012), p. 103.
120. Guy Bradley, *Early Rome to 290 BC: The Beginnings of the City and the Rise of the Republic*, p. 327.

Chapter 6
1. Michael J. Dobson, *The Army of the Roman Republic: The Second Century BC, Polybius and the Camps at Numantia, Spain* (Oxford, 2008), p. 47.
2. Livy 8.8.
3. Dion. Hal. *Ant. Rom.* 5.75.4ff. Kubitschek, 'Legio. Republikanische Zeit', col. 1202f.
4. Jeremy Armstrong, 'Organised chaos: *manipli, socii* and the Roman army c.300', in Jeremy Armstrong and Michael P. Fronda, eds., *Romans at War: Soldiers, Citizens, and Society in the Roman Republic* (London, 2020), pp. 84–9.
5. Echeverría Rey, 'Weapons, Technological Determinism and Ancient Warfare', p. 30 fn 21.
6. Pol. 1.1.5, i.e., 9.9.9. For Polybius' audience, see F.W. Walbank, *Polybius* (Berkeley, 1972), pp. 3–6., Erich S. Gruen, 'Polybius and Josephus on Rome', in F. W. Walbank, Bruce Gibson and Thomas Harrison, eds., *Polybius and His World: Essays in Memory of F.W. Walbank* (2013), p. 256f and Peter Derow, Andrew Erskine and Josephine Crawley Quinn, eds., *Rome, Polybius, and the East* (Oxford, 2015), p. 6.
7. Contra: M. Dubuisson, *Le latin de Polybe : les implications historiques d'un cas de bilinguisme* (Paris, 1985), pp. 266, 269. For Polybius' view of the Romans, see Andrew Erskine, 'How to rule the world: Polybius Book 6 reconsidered', in Bruce John Gibson and Thomas Harrison, eds., *Polybius and His World: Essays in Memory of F.W. Walbank* (Oxford ; New York, 2013), pp. 239–245.
8. Pol. 5.90.8; 6.57.10; 2.35.2f; 9.11.2.; 9.24–25.
9. Pol. 31.22.8–11, 2.15.6, 3.87.6 and a translated treaty: 3.22.3.
10. Armstrong, 'Organised chaos: *Manipuli, socii,* and the Roman army c.300', p. 82. Erdkamp notes Polybius' lack of mention of Italian allied troops, seeing them as

essentially Romans: Paul Erdkamp, 'Polybius and Livy on the allies in the Roman army', in Lukas De Blois, Elio Lo Cascio, Olivier Hekster and Gerda De Kleijn, eds., *Impact of the Roman Army (200 BC-AD 476): Economic, Social, Political, Religious, and Cultural Aspects: Proceedings of the Sixth Workshop of the International Network Impact of Empire (Roman Empire, 200 B.C.-A.D. 476), Capri, March 29-April 2, 2005* (Boston (Mass.), 2007), pp. 49, 55.

11. One Polybius' military knowlegde, see Walbank, *Polybius*, pp. 10f, 33, 88–9 and Burns, 'The Homogenisation of Military Equipment under the Roman Republic', p. 70.
12. Pol. 1.36.5; 1.39.10; 1.42; 1.47f; 1.59.6; 7.13; 8.12.6.
13. On Polybius' early life, see F. W. Walbank, *A Historical Commentary on Polybius, Volume 1* (Oxford, 1957), p. 1f.
14. Pol. 3.58; 13.3.
15. Elizabeth H. Pearson, *Exploring the Mid-Republican Origins of Roman Military Administration: With Stylus and Spear* (London, 2021), p. 15.
16. Armstrong, 'Organised chaos: *manipli, socii* and the Roman army c.300', pp. 82, 91.
17. Joseph. *B.J.* 5.460.
18. Sen. *QNat.* 2.26.6; 30.1; 6.22.2; 6.17.3; 5.15.1.
19. Pol. 6.19.6. There is a modern debate on the length of service expected of Roman citizens: Pearson, *Exploring the Mid-Republican Origins of Roman Military Administration: With Stylus and Spear*, p. 18f.
20. Nicholas Sekunda, 'Military Forces. A. Land forces B. Naval forces', in Philip Sabin, Hans van Wees and Michael J. Whitby, eds., *The Cambridge History of Greek and Roman Warfare: Volume I Greece, the Hellenistic World and the Rise of Rome* (Cambridge, 2007), p. 334.
21. Livy 42.34.
22. Pol. 6.21.4. Erdkamp, 'Polybius and Livy on the allies in the Roman army', p. 55.
23. Pol. 6.20.
24. Contra Keppie, *The Making of the Roman Army*, p. 35, who claims, without evidence, that their assignment was for administrative purposes.
25. Dobson, *The Army of the Roman Republic: The Second Century BC, Polybius and the Camps at Numantia, Spain*, pp. 48, 63, 84.
26. Contra Sage, *The Army of the Roman Republic: From the Regal Period to the Army of Julius Caesar*, p. 159 but acknowledges the fighting hand to hand in Livy 31.35.5–6.
27. Pol. 6.21.9. Total numbers: 6.20.9
28. Pol. 6.20.9, 1.16.2, 2.24.13. For a discussion of legionary numbers, see Dobson, *The Army of the Roman Republic: The Second Century BC, Polybius and the Camps at Numantia, Spain*, p. 49f.
29. Pol. 6.20.8, 3.107.11.
30. Pol. 6.21.7.
31. For example: Dobson, *The Army of the Roman Republic: The Second Century BC, Polybius and the Camps at Numantia, Spain, Ibid.*, M.C. Bishop and J.C.N. Coulston, *Roman Military Equipment From the Punic Wars to the Fall of Rome* (Oxford, 2006), p. 50f., Jana Horvat, 'The Hoard of Roman Republican Weapons from Grad near Šmihel', *Arheološki Vestnik*, 53 (2002) p. 138 and Duncan B. Campbell, 'ca', in Brian Campbell and Lawrence A. Tritle, eds., *The Oxford Handbook of Warfare in the Classical World* (Oxford, 2013), pp. 5–7.

32. Pol. 6.22.10. See also modern reconstructions of Roman javelins in Klejnowski Grzegorz, 'Hasta Velitaris: the first edge of the Roman army', *Res Militaris Studia nad wojskowością antyczną*, II (2015).
33. Pol. 6.22. See also Aelian 2 and 17 in Aelian, *The Tactics of Aelian or On the Military Arrangements of the Greeks* (Barnsley, 2012) and Arr. *Tact*.14.
34. Pol. 6.23. Campbell, 'Arming the Romans for Battle', pp. 7–8. For an unconvinced modern view, see Hildinger, *Swords Against the Senate: The Rise of the Roman Army and the Fall of the Republic*, p. 24. Gilliver suggests that great uniformity was unlikely: C.M Gilliver, 'Display in Roman Warfare: The Appearance of Armies and Individuals in the Battlefield', *War in History*, 14 (2007), pp. 3–4.
35. Pol. 6.23.6. Campbell, 'Arming the Romans for Battle', Dobson, *The Army of the Roman Republic: The Second Century BC, Polybius and the Camps at Numantia, Spain*, pp. 4–5.
36. Feugère, *Les armes des romains de la république à l'antiquité tardive*, p. 98. Campbell, 'Arming the Romans for Battle', p. 4.
37. Pol. 15.15.8.
38. Echeverría Rey, 'Weapons, Technological Determinism and Ancient Warfare', p. 27.
39. Ascl. *Tact*. 2.8.
40. s.v. Optio in C.T. Lewis, C. Short, E.A. Andrews and W. Freund, *A Latin Dictionary: Founded on Andrews' Edition of Freund's Latin Dictionary* (Oxford, 1945).
41. Pol. 6.24.
42. Pol. 6.25.
43. Pol. 6.26.5–6
44. Pol. 6.26.8–9.
45. Dobson, *The Army of the Roman Republic: The Second Century BC, Polybius and the Camps at Numantia, Spain*, p. 51.
46. Pol. 18.18.2f.
47. Pol. 6.31.10 - 6.41.
48. Pol. 6.42.
49. Pol. 18.18.6–18
50. Dobson, *The Army of the Roman Republic: The Second Century BC, Polybius and the Camps at Numantia, Spain*, p. 68f.
51. *Ibid.*
52. Pol. 1.6.6.
53. Pol. 1.65.7; 1.67.4f; 1.81.10.
54. Pol. 3.33.8; 3.62.2f; 3.76.13; 9.3.9.
55. Pol. 9.12–16.
56. Pol. 3.115.3; 11.21.4.
57. Pol. 3.117.5.
58. Pol. 12.18.3
59. Pol. 5.2.6.
60. Pol. 5.98.1; 9.19.5.
61. Pol. 10.15–17.5 but 11.23.11.
62. Pol. 1.2; 1.37.4; 1.63.6; 1.84.6f; 3.28.1f.
63. Pol. 30.22; 39.2.1.
64. Pol. 15.9.7.
65. Pol. 18.24.10.

66. Pol. 15.12.8.
67. Pol. 15.13.7.
68. Pol. 15.14.3.
69. Pol. 18.28.1–6.
70. Pol. 18.29.6–11.
71. Pol. 18.28.12.
72. Pol. 18.29 - 18.30.1–4.
73. Pol. 29.17.
74. Pol. 18.30. For a discussion of frontages, see Michael J. Taylor, 'Visual evidence for Roman infantry tactics', *Memoirs of the American Academy in Rome*, 59 (2014), pp. 106–9 and Goldsworthy, *The Roman Army at War: 100 BC–AD 200*, p. 179.
75. Pol. 18.31.
76. Armstrong, 'Organised chaos: *manipli, socii* and the Roman army c.300', p. 83.
77. Sanz, 'Not so different: individual fighting techniques and battle tactics of Roman and Iberian armies within the framework of warfare in the Hellenistic Age', pp. 7–8.
78. Taylor, 'Visual evidence for Roman infantry tactics', pp. 116–17.
79. Armstrong, 'Organised chaos: *manipli, socii* and the Roman army c.300', pp. 96–8. See also Marian Helm, 'Poor man's war – rich man's fight. Military integration in Republican Rome', in Jeremy Armstrong and Michael P. Fronda, eds., *Romans at War: Soldiers, Citizens, and Society in the Roman Republic* (London, 2020), p. 106f.
80. Sekunda, 'Military Forces. A. Land forces B. Naval forces', p. 349.
81. Angelos Chariotis, 'The Impact of War on the Economy of Hellenistic *Poleis*: Demand Creation, Short-Term Influences and Long-Term Impacts', in Zosia H. Archibald, John K. Davies and Vincent. Gabrielsen, eds., *The Economies of Hellenistic Societies, Third to First Centuries BC* (Oxford, 2011), p. 124.
82. Sekunda, 'Military Forces. A. Land forces B. Naval forces', pp. 334–5.
83. For Roman population, see Brunt, *Italian Manpower*, p. Chs. IV, XI.
84. Pol. 1.26.9, 2.24.
85. Sekunda, 'Military Forces. A. Land forces B. Naval forces', pp. 336, 348.
86. Pol. 30.25.3–11.
87. Pol. 11.23.1, 11.33.1.
88. In 210 BCE: Livy 25.39.1. His earlier reference in Book 2 may be assumed to be anachronistic: Dobson, *The Army of the Roman Republic: The Second Century BC, Polybius and the Camps at Numantia, Spain*, p. 58.
89. Bell, 'Tactical Reform in the Roman Republican Army', p. 405.
90. Dobson, *The Army of the Roman Republic: The Second Century BC, Polybius and the Camps at Numantia, Spain*, pp. 59, 414.
91. Sall. *Iug.* 49.6.
92. Dobson, *The Army of the Roman Republic: The Second Century BC, Polybius and the Camps at Numantia, Spain*, p. 62f.
93. Pol. 11.23.1.
94. See Bell, 'Tactical Reform in the Roman Republican Army', p. 414.
95. Taylor, 'Visual evidence for Roman infantry tactics', p. 105, fn 11.
96. Regarding food supplies, see Paul Erdkamp, 'The Corn Supply of the Roman Armies during the Third and Second Centuries B.C', *Historia: Zeitschrift für alte Geschichte*, 44 (1995), p. 172f.
97. Sanz, 'Not so different: individual fighting techniques and battle tactics of Roman and Iberian armies within the framework of warfare in the Hellenistic Age', p. 3f.

For an example of speculation, see Hildinger, *Swords Against the Senate: The Rise of the Roman Army and the Fall of the Republic*, pp. 25–29. Studies of Roman battles in the later republican and early imperial period are useful and will be discussed in the next chapter: Goldsworthy, *The Roman Army at War: 100 BC–AD 200*.

98. Pol. 6.40.
99. Alexander Zhmodikov, 'Roman Republican Heavy Infantrymen in Battle (IV-II Centuries B.C.)', *Historia: Zeitschrift für alte Geschichte*, 49 (2000), p. 78. but contra Goldsworthy, *The Roman Army at War: 100 BC–AD 200*, p. 33. Zhmodikov strangely argues that there is no evidence that Polybius had actually seen Romans in battle and that, as a result, he underplays the importance of missile weapons in Roman battle although Polybius is clear on the number of missile-armed troops in a legion.
100. Sanz, 'Not so different: individual fighting techniques and battle tactics of Roman and Iberian armies within the framework of warfare in the Hellenistic Age', pp. 4–6. Jeremiah McCall, 'The manipular army system and command decisions in the second century', in Jeremy Armstrong and Michael P. Fronda, eds., *Romans at War: Soldiers, Citizens, and Society in the Roman Republic* (London, 2020), p. 223f.
101. Philip Sabin, 'The Face of Roman Battle', *The Journal of Roman Studies*, 90 (2000), p. 7.
102. E.g., Pol 15.24; Livy 30.34.9–12; Caes. *BC.* 1.45, 3.94.
103. Sanz, 'Not so different: individual fighting techniques and battle tactics of Roman and Iberian armies within the framework of warfare in the Hellenistic Age', p. 7 and fn 9.
104. Michael J Taylor, 'Roman infantry tactics in the mid-Republic: a reassessment', Historia: Zeitschrift für alte Geschichte, (2014), pp. 301–22.
105. Brunt, *Italian Manpower*, p. 625–34.
106. Pearson, *Exploring the Mid-Republican Origins of Roman Military Administration: With Stylus and Spear*, p. Ch. 1. For a critique of Polybius' description, see Brunt, *Italian Manpower*, p. 625–34. On the meaning of *dilectus*, see Ibid., p. 635–8.
107. Plut. *Vit. Marc.*13.
108. Livy 27.38.
109. Pol. 6.19.
110. Livy 27.46.
111. Plut. *Vit. Marc.* 12.
112. Brunt, *Italian Manpower*, p. 671–6.
113. Pearson, *Exploring the Mid-Republican Origins of Roman Military Administration: With Stylus and Spear*, p. 186–7, Ch. 2.
114. Pol. 2.24. Forsythe, *A Critical History of Early Rome*, p. 366.
115. For example, see Victor Davis Hanson, *The Western Way of War: Infantry Battle in Classical Greece* (New York, 1989), p. Ch 2.
116. *Ibid.*, Preface.
117. Pol.10.12.4–7; J.E. Lendon, 'The Rhetoric of Combat: Greek Theory and Roman Culture in Caesar's Battle Descriptions', *Classical Antiquity*, 18 (1999), p. 295.
118. Plautus. *Amph.* 188–262, *Asin.* 15, *Capt.* 67–8, *Cas.* 87–8, *Cist.* 197–202, i.e., W.V. Harris, *War and Imperialism in Republican Rome, 327–70 B.C* (Oxford, 1979), p. 43.
119. *Ibid.*, pp. 102–3, 231 n.2.
120. Joëlle Napoli, 'Rome et le recrutement de mercenaires', *Revue historique des armées*, (2010), p. 2.

121. Livy. 25.37
122. The details of his appointment are unclear. See comments in A. Goldsworthy, *The Punic Wars* (London, 2000), pp. 269–70.
123. Livy 29.1; Pol. 3.106.7. On the Senate's allocation, see the discussion in E. Badian, *Publicans and Sinners* (Oxford, 1972), p. 21. *Publicani* provided credit to fund supplies: J.S. Richardson, *Hispaniae: Spain and the Development of Roman Imperialism, 218–82 B.C.* (Cambridge, 1986), p. 39. There was evidence of corruption later: Livy 25.3.9–11.
124. Livy 26.17, 19.
125. Livy 27.10. Senate sent supplies to P. Scipio in 217: Livy 22.22.1. A Carthaginian squadron had captured another shipment soon after it left Ostia: Livy 22.11.6.
126. Richardson, *Hispaniae: Spain and the Development of Roman Imperialism, 218–82 B.C.*, p. 58.
127. Regarding the allies, see Scipio's treatment of Mandonius: Livy 28.34 and App. *Hisp* 15. For the role of booty, see: Livy 28.24. The army commanded by Scipio's father and uncle was supplied from Italy: Livy 23.48. On the allied contributions to support the Roman forces, see Livy 29.3.5 and Pol. 11.25.9.
128. Pol. 9.24; Livy 28.12; Plut. *Vit. Fab.* 2.5.
129. Livy 26.47.
130. Livy 26.51; Pol. 10.20.
131. Pol. 2.30, 2.33, 3.11, 3.2–3, 4.23.6–7.
132. Livy 27.17.
133. Pol 10.9–16, 10.35.5.
134. Livy 28.17.
135. Livy 27.18; Pol. 10.39.
136. Livy 28.13f; Pol. 9.22f.
137. Livy 28.24.
138. On Scipio's relations with Spanish leaders, see: E. Badian, *Foreign Clientelae* (Oxford, 1958), pp. 117–18.
139. Livy 28.29.
140. App. *Hisp.* 38.
141. Pol. 11.33.7.
142. On Scipio's consulship in 205, see: H.H. Scullard, *Roman Politics 220–150 B.C.* (Oxford, 1973), p. 75–76.
143. Edward Dale Clark, 'Roman legionary forces in Sicily during the Second Punic War: the number of legions stationed on the island from 214 to 210 B. C.', *The Ancient History Bulletin*, 8 (1994), p. 133–140. Brunt, *Italian Manpower*, p. 420.
144. Livy 28.45.
145. Livy 29.1. App. *Hisp.* 8.
146. Livy 29.24.
147. App. *BCiv.* 55.
148. Plut. *Vit. Cat. Mai.* 3.
149. Livy 29.22.
150. Livy 29.1.
151. Livy 29.25.
152. Livy 30.7–8; Pol. 14.4–5; App. Pun. 21. See Goldsworthy, *The Punic Wars*, p. 293.
153. Livy 30.8.
154. Pol. 14.8.

155. Goldsworthy, *The Punic Wars*, p. 296.
156. Livy 30.33; Pol. 15.9.
157. App. *Pun.* 41.
158. Livy 30.34. Pol. 15.14. Appian simply says that Scipio formed a stronger battle line: App. Pun. 47. Goldsworthy believes that Scipio executed a double envelopment: Goldsworthy, *The Punic Wars*, pp. 306–7.
159. Pol. 11.1; Livy 28.48.
160. Livy 33.9.
161. Pol. 18.32.
162. Goldsworthy, *The Punic Wars*, p. 319. Note the enlistment of 'volunteers' from the African army: Livy 31.8.

Chapter 7

1. Judson, *Caesar's Army; a Study of the Military Art of the Romans in the Last Days of the Republic*. Dodge, *Caesar: A History of the Art of War Among the Romans Down to the End of the Roman Empire*.
2. Cary, *A History of Rome down to the Reign of Constantine*, p. 308; H.H. Scullard, *From the Gracchi to Nero* (London, 1959), p. 58; Bell, 'Tactical Reform in the Roman Republican Army', p. 405f; Keppie, *The Making of the Roman Army*, p. 63f.
3. Goldsworthy, *The Roman Army at War: 100 BC–AD 200*, p. 34.
4. Burns, 'The Homogenisation of Military Equipment under the Roman Republic', p. 71.
5. On the Marian army reforms, Bell credits Marquardt with the theory of the Marian Army Reforms: Bell, 'Tactical Reform in the Roman Republican Army', p. 404. See also Kromayer and Veith, *Heerwesen und Kriegsführung der Griechen und Römer*, p. 377; Cary, *A History of Rome down to the Reign of Constantine*, p. 297, Parker, *The Roman Legions*, p. 27–44, Delbrück, *Warfare in Antiquity. History of the Art of War*, p. 415, Keppie, *The Making of the Roman Army*, p. 63f, M. Junkelmann, *Die Legionen des Augustus* (Mainz am Rhein, 1986), p. 87, Webster, *The Roman Imperial Army*, p. 21, Cagniart, 'The Late Republican Army (146–30BC)', p. 86–87, H.H. Scullard, *From the Gracchi to Nero* (London, 2010), p. 58. Goldsworthy acknowledges that the changes claimed for Marius had been happening for some time: Goldsworthy, *Roman Warfare*, p. 106.
6. Cadiou, *L'armée imaginaire: les soldats prolétaires dans les légions Romaines au dernier siécle de la République*, p. 10.
7. *Ibid.*
8. *Ibid.*, p. 58.
9. Delbrück, *Warfare in Antiquity: History of the Art of War*, pp. 377–8.
10. Kromayer and Veith, *Heerwesen und Kriegsführung der Griechen und Römer*, p. 376.
11. Parker, *The Roman Legions*, p. 28.
12. Webster, *The Roman Imperial Army*, p. 22.
13. Keppie, *The Making of the Roman Army*, pp. 63–4.
14. Bell, 'Tactical Reform in the Roman Republican Army', p. 404.
15. Gilliver and Kertész accept Bell's argument that cohorts originated in Roman warfare in Spain: *Ibid.*, p. 410; Gilliver, *The Roman Art of War*, p. 18. I. Kertész, 'The Roman Cohort Tactics – Problems of Development', *Oikumene*, 1 (1976), p. 93.
16. Bell, 'Tactical Reform in the Roman Republican Army', p. 422.

17. *Ibid.*, p. 417f. The problem of the timing of the change is described anachronistically in these terms: 'It is characteristic of all armies that reforms desirable on military grounds are liable to be delayed or perverted for non-military reasons'. While the comment may have some relevance to modern states, it is anachronistic and misleading to apply it to the Romans.
18. Kertész, 'The Roman Cohort Tactics – Problems of Development', p. 89. Gilliver, *The Roman Art of War*, p. 18.
19. Michael J. Taylor, 'Tactical Reform in the Late Roman Republic: The View from Italy', *Historia*, 68 (2019), pp. 76–94.
20. Sall. *Iug.* 86; Plut. *Vit. Mar.* 9.1; Flor. 1.36.13; Val. Max. 2.3.1; Aul. Gel. *NA.* 16.10.14; Ps-Quint. *Decl. Mai.* 3.5; Pliny *N. H.* 10.16; Exsuperantius 9, probably using Sallust's Histories. On the other hand, Velleius does not mention Marius' enlistment of *capite censi*: Vell. Pat. 2.2.
21. M. Crawford, *The Roman Republic* (Cambridge, Mass., 1992), p. 125. Enn. *Ann.* 183 V, in reference to the Pyrrhic War.
22. Cary, *A History of Rome down to the Reign of Constantine*, p. 308.
23. *Ibid.*, p. 303.
24. Parker, *The Roman Legions*, p. 25.
25. Smith, *Service in the Post-Marian Roman Army*, p. 11f.
26. *Ibid.*, p. 11.
27. Gabba, *Republican Rome, the Army and the Allies*, pp. 25–6, 33, 34–5, 52.
28. Harmand, *L'Armée et le soldat à Rome de 107 à 50 avant notre ère*, p. 11.
29. Webster, *The Roman Imperial Army*, pp. 15, 19.
30. *Ibid.*, p. 18.
31. *Ibid.*, p. 19.
32. *Ibid.*, pp. 20–22.
33. Brunt, *Italian Manpower*, p. 637.
34. Gabba, *Republican Rome, the Army and the Allies*, pp. 25–6, 33–5, 52. P. A. Brunt, 'The army and the land in the Roman revolution', *The Journal of Roman Studies*, LII (1962), p. 256, rejecting Gabba.
35. Livy 10.33.1.
36. Livy 27.18.10, 25.39.1. viz. Frontin. *Str.* 2.6.2; Livy 34.12.6, 34.14.1, 34.14.7, 34.14.10, 34.15.1, 34.19.9, 34.20.3.
37. Livy 25.39.1, 27.49.4.
38. Pol. 11.23.1; Livy 28.14.7. For a detailed discussion, see Kertész, 'The Roman Cohort Tactics - Problems of Development', p. 90f.
39. *Ibid.*, p. 95.
40. Parker, *The Roman Legions*, p. 28f. See the comments in Bell, 'Tactical Reform in the Roman Republican Army', p. 405.
41. Kromayer and Veith, *Heerwesen und Kriegsführung der Griechen und Römer*, p. 378. Delbrück, *Warfare in Antiquity: History of the Art of War*, p. 415. Bell, 'Tactical Reform in the Roman Republican Army', p. 419f. Cary, *A History of Rome down to the Reign of Constantine*, p. 306. Scullard, *From the Gracchi to Nero* (1976), p. 58. Scullard, *From the Gracchi to Nero* (2010), p. 47. By implication: Parker, *The Roman Legions*, p. 43, Webster, *The Roman Imperial Army*, p. 21 and Keppie, *The Making of the Roman Army*, p. 64.
42. Sall. *Iug.* 105.2 archers and slingers, 49.6.

43. Frontin. *Str.* 2.3.17.
44. Bell, 'Tactical Reform in the Roman Republican Army', p. 422.
45. *Ibid.* 419.
46. Kromayer and Veith, *Heerwesen und Kriegsführung der Griechen und Römer*, p. 378. Parker, *The Roman Legions*, p. 43. K.R. Dixon and P. Southern, *The Roman Cavalry* (London, 1992), p. 22 implied.
47. Pol. 35.4.3ff
48. Sall. *Iug.* 95.1.
49. Sall. *Iug.* 98.1.
50. Val. Max. 5.8.4.
51. Plut. *Vit. Pomp.* 22. The continued state issue of horses to wealthy young men neither proves nor disproves that the young men concerned actually served as cavalry. What it does show is that the practice of issuing public horses had not died out in Rome, at least. We have no idea what provision was made for cavalrymen in the former Italian allied states. See also the list of foreign troops in Caes. *BCiv.* 3.3–4. This list is suspect. It may simply be a catalogue of eastern exotica designed to incriminate his enemies.
52. Plut. *Vit. Pomp.* 64.
53. Caes. *BCiv.* 3.93. Jeremiah B. McCall, *The Cavalry of the Roman Republic: Cavalry Combat and Elite Reputations in the Middle and Late Republic* (London, 2002), p. 102.
54. It must be admitted that the cavalry could be those mentioned in Caes. *BCiv.* 3, 4. Saddington believes that Roman cavalry served under Pompey at Pharsalus and at Munda: D.B. Saddington, *The Development of the Auxiliary Forces from Caesar to Vespasian* (Harare, 1982), p. 25.
55. App. *BCiv.* 5.138.
56. *Tabula Heracleensis* (ILS 6085), 88. On the date, see Michael H. Crawford and J. D. Cloud, *Roman Statutes* (London, 1996), p. 360.
57. Keppie is aware of some of the evidence but he is unwilling to abandon the assumption that Roman cavalry were abolished: Keppie, *The Making of the Roman Army*, p. 79. On the final separation of the cavalry from the infantry, see P. Brennan, T. W. Hillard, R.A. Kearsley, C.V.E. Nixon and A.M. Nobbs, 'Divide and Fall: the separation of legionary cavalry and the fragmentation of the Roman Empire', *Ancient History in a Modern University* (2, 1998), pp. 238–44.
58. John the Lydian credits Marius with the creation of the legions: Lydus *Mens.* 16.3. John also credits Marius with the institution of *tirones*, recruits: Lydus *Mens.* 48.1. It is difficult to have much faith in John on these matters though it raises questions as to what John had before him.
59. Plut. *Vit. Mar.* 25. The use of wooden pins in the *pila* did not seem to survive Marius. There is no known evidence of their use later.
60. Sall. *Iug.* 84.3.
61. Sall. *Iug.* 86.3.
62. Smith, *Service in the Post-Marian Roman Army*, p. 10; C. Nicolet, *The World of the Citizen in Republican Rome* (London, 1980), p. 93.
63. François Gauthier, 'The transformation of the Roman army in the last decades of the Roman Republic', in Jeremy Armstrong and Michael P. Fronda, eds., *Romans at War: Soldiers, Citizens, and Society in the Roman Republic* (London, 2020), p. 283.
64. Walbank, *A Historical Commentary on Polybius, Volume 1*, p. 6.

65. Velleius Paterculus, an experienced, prosaic soldier and supporter of the Emperor Tiberius, was of the opinion that the Italian allies were justified in revolt because the allied states supplied twice the number of infantry and cavalry as the Romans: Vell. Pat. 2.15.2.
66. For this, see W. Schmitthenner, 'The Armies of the Triumval Period: A Study of the Origins of the Roman Imperial Legions', (D.Phil. thesis, Oxford, 1958).
67. Suet. *Iul.* 56. A Goldsworthy, 'Instinctive Genius. The Depiction of Caesar the General', in A. Powell and K. Welch, eds., *Julius Caesar as Artful Reporter* (London, 1998), pp. 85, 102, 194–5. R.M. Rambaud, *L'art de la déformation historique dans les commentaires de César* (Paris, 1966), p. 245.
68. Smith, *Service in the Post-Marian Roman Army*, p. 44f.
69. Cic. *De Off.* 1, 25; Plut. *Vit. Crass.* 2. Brutus provided his soldiers with armour decorated in gold and silver to improve their loyalty: Plut. *Vit. Brut.* 38.3. Caesar did the same: Suet. *Iul.* 67.
70. Harmand, *L'Armée et le soldat à Rome de 107 à 50 avant notre ère*, p. 29.
71. *Ibid.*, p. 26.
72. Brunt, *Italian Manpower*, p. 687.
73. *Ibid.*, p. 693.
74. App. *BCiv.* 3.9.68.
75. App. *BCiv.* 3.9.69.
76. Caes. *BG.* 8.8. Goldsworthy, *The Roman Army at War: 100 BC–AD 200*, p. 25.
77. Caes. *BAfr.* 45
78. Caes. *BAfr.* 85; Cic. *Fam.* 10.30.
79. E.g., Caes. *BG.* 6.40; Cic. *Fam.* 10.24.3.
80. Cic. *Fam.* 10.32.4.
81. Val. Max. 2.3.3.
82. App. *BCiv.* 3.7.43.
83. Saddington, *The Development of the Auxiliary Forces from Caesar to Vespasian*, p. 25. Goldsworthy, *The Roman Army at War: 100 BC–AD 200*, p. 35.
84. Suet. *Iul.* 24. Smith, *Service in the Post-Marian Roman Army*, p. 57.
85. Caes. *BCiv.* 2.20, 1.87; *BHisp.* 7.
86. Harmand, *L'Armée et le soldat à Rome de 107 à 50 avant notre ère*, p. 34.
87. Caes. *BAlex.* 50, 53.
88. Cic. *Att.* 5.15, 5.18.
89. App. *BCiv.* 4.10.75.
90. Caes. *BAlex. 34*
91. Viz. Cic. *Att.* 6.1.
92. Caes. *BCiv.* 2.34–35.
93. E.g., App. *BCiv.* 1.59, 74, 106.
94. E.g., App. *BCiv.* 1.42, 85, 87, 88, 90, 91; 2.38; App. *Mith.* 12.59
95. Sall. *Iug.* 41.4, 94.6.
96. Sall. *Cat.* 11.5f; App. *Mith.* 12.82f; App. *BCiv.* 4.3, 5.124; Plut. *Vit. Luc.* 17.5, 19.3; Plut. *Vit. Pomp.* 11, 3–4; Plut. *Vit. Brut.* 46.1; Plut. *Vit. Ant.* 48.2; Dio Cass. 36, 45.12.1f; Caes. *BC.* 3.31.
97. App. *BCiv.* 2.30, 47.
98. E.g. App. *BCiv.* 2.67; Plut. *Vit. Luc.* 14.2.
99. App. *BCiv.* 5.2.17.

100. Lee L. Brice, '*SPQR* SNAFU: indiscipline and internal conflict in the late Republic', in Jeremy Armstrong and Michael P. Fronda, eds., *Romans at War: Soldiers, Citizens, and Society in the Roman Republic* (London, 2020), p. 253.
101. *Ibid.*, p. 257.
102. Sall. *Iug.* 46.7, 49.6.
103. Dobson, *The Army of the Roman Republic: The Second Century BC, Polybius and the Camps at Numantia, Spain*, p. 63.
104. Plut. *Vit. C. Gracch.* 5.1; Asc. 68 C.
105. Junkelmann, *Die Legionen des Augustus*, p. 100.
106. *Auxilia* is a contraction of the term *auxila sociorum*, i.e., help of the allies, Caes. *BG* 8.17. On the distinction between Roman an allied troops, see *aut legiones et auxilia*: 8.25.
107. Cic. *Fam.* 15.1.5.
108. E.g., Cic. *Fam.* 15.4.8; Caes. *BCiv.* 1.42.1.
109. Caes. *BG.* 2.19, 6.5.
110. Caes. *BCiv.* 1.27.
111. Caes. *BCiv.* 1.43–6; 3.75.
112. Goldsworthy, *The Roman Army at War: 100 BC–AD 200*, p. 18.
113. Goldsworthy, *Roman Warfare*, p. 108. Cagniart, 'The Late Republican Army (146–30BC)', p. 85. Bell, 'Tactical Reform in the Roman Republican Army'. Contra: François Cadiou, 'Les guerres en *Hispania* et l'émergence de la cohorte légionnaire dans l'armée romaine sous la république: Une révision critique', *Gladius*, XXI (2001), pp. 167–82.
114. Goldsworthy, *The Roman Army at War: 100 BC–AD 200*, p. 34.
115. Sall. *Iug.* 95.2: *cohors Paeligna cum velitaribus armis*, i.e., a Paelignian cohors with light arms.
116. Caes. *BGal.* 6.34, 7.47.7, 7.50.4; Cic. *Phil.* 5.12, *Att.* 9.10.2; Plaut. *Mostell.* 312, 1048.
117. For a comprehensive discussion of the evidence, see Dobson, *The Army of the Roman Republic: The Second Century BC, Polybius and the Camps at Numantia, Spain*, p. 58f.
118. Keppie, *The Making of the Roman Army*, p. 64.
119. Rüstow, *Heerwesen und Kriegführung C. Julius Cäsars*, p. 5f.
120. Dobson, *The Army of the Roman Republic: The Second Century BC, Polybius and the Camps at Numantia, Spain*, p. 62. Cagniart, 'The Late Republican Army (146–30BC)', p. 89.
121. Saddington, *The Development of the Auxiliary Forces from Caesar to Vespasian*, p. 25.
122. McCall, *The Cavalry of the Roman Republic: Cavalry Combat and Elite Reputations in the Middle and Late Republic*, p. 107f.
123. Pol. 6.20.9
124. Pol. 6.26.7.
125. McCall, *The Cavalry of the Roman Republic: Cavalry Combat and Elite Reputations in the Middle and Late Republic*, p. 112.
126. Feugère, *Les armes des romains de la république à l'antiquité tardive*, pp. 101–102. Bishop and Coulston, *Roman Military Equipment from the Punic Wars to the Fall of Rome*, pp. 50–53. Raffaele. D'Amato and Graham. Sumner, *Arms and Armour of the Imperial Roman Soldier: From Marius to Commodus, 112 BC–AD 192* (London, 2009), p. 7.
127. Plut. *Vit. Mar.* 25.
128. Christopher Matthew, 'The Battle of Vercellae and the Alteration of the Heavy Javelin (Pilum) by Gaius Marius - 101 BC', *Antichthon*, 44 (2010), p. 63.

129. Caes. *BG*. 1.52.
130. Caes. *BG* 1.25.
131. Caes. *BG* 2.27.
132. Caes. *BAfr.* 16.
133. Feugère, *Les armes des romains de la république à l'antiquité tardive*, p. 98. Bishop and Coulston, *Roman Military Equipment from the Punic Wars to the Fall of Rome*, pp. 54–56. D'Amato and Sumner, *Arms and Armour of the Imperial Roman Soldier: From Marius to Commodus, 112 BC-AD 192*, p. 14.
134. Bishop and Coulston, *Roman Military Equipment from the Punic Wars to the Fall of Rome*, pp. 61–2. Feugère, *Les armes des romains de la république à l'antiquité tardive*, p. 93.
135. Caes. *BG*. 2.21.
136. Bishop and Coulston, *Roman Military Equipment from the Punic Wars to the Fall of Rome*, p. 66. Feugère, *Les armes des romains de la république à l'antiquité tardive*, pp. 84–86. D'Amato and Sumner, *Arms and Armour of the Imperial Roman Soldier: From Marius to Commodus, 112 BC-AD 192*, p. 34f.
137. Caes. *BAfr.* 12.
138. D'Amato and Sumner, *Arms and Armour of the Imperial Roman Soldier: From Marius to Commodus, 112 BC-AD 192*, p. 18, Figure 8.
139. Michael J Taylor, 'The battle scene on Aemilius Paullus's Pydna monument: a reevaluation', *Hesperia: The Journal of the American School of Classical Studies at Athens*, 85 (2016), p. 569.
140. Bishop and Coulston, *Roman Military Equipment From the Punic Wars to the Fall of Rome*, p. 63.
141. D'Amato and Sumner, *Arms and Armour of the Imperial Roman Soldier: From Marius to Commodus, 112 BC-AD 192*, pp. 38–39. David Sim, 'The manufacture of disposable weapons for the Roman army', *Journal of Roman Military Equipment*, 3 (1992), p. 123.
142. David Sim and Jaime Kaminski, *Roman Imperial Armour: The Production of Early Imperial Military Armour* (Oxford, 2012), p. 132.
143. Campbell, 'Arming the Romans for Battle', p. 7.
144. Bishop and Coulston, *Roman Military Equipment From the Punic Wars to the Fall of Rome*, p. 63.
145. Slaves: Caes. *BAfr.* 47, *BG*. 8.10; *BAlex.* 73, *BC*. 3.6. Suet. *Iul.* 26, 38.
146. Caes. *BCiv.* 3.44.
147. Claudio. Franzoni, *Habitus atque habitudo militis: monumenti funerari di militari nella Cisalpina Romana* (Roma, 1987), p. 26. L. Keppie, 'A centurion of legio Martia at Padova?', *Journal of Roman Military Equipment Studies*, 2 pp. 115–121. For images, see http://lupa.at/14644.
148. Pol. 6.24.
149. Caes. *BG*. 1.40.
150. Caes. *BG*. 5.39, 5.44.
151. Caes. *BCiv.* 1.20, 1.13.
152. Caes. *BHisp.* 23.
153. Caes. *BG*. 6.40
154 Cic. *Att.* 5.20.
155. Goldsworthy, *The Roman Army at War: 100 BC-AD 200*, p. 34f, Ch. 5.
156. Caes. *BG*. 1.24.

157. Caes. BCiv. 1.83.
158. Caes. *BG.* 2.25
159. Caes. *BC.* 3.92; Lendon, 'The Rhetoric of Combat: Greek Theory and Roman Culture in Caesar's Battle Descriptions', p. 280.
160. Caes. *BG.* 8.14.
161. Caes. *BAfr.* 29.
162. Pol. 6.24.6.
163. Caes. *BG.* 4.26.
164. Caes. *BCiv.* 1.71.
165. Caes. *BG.* 2.20.
166. Caes. *BAfr.* 15
167. Pol. 6.21.
168. Livy 22.38.
169. Dion. Hal. *Ant. Rom.* 10.18.2.
170. Caes. *BG.* 3.4.
171. Caes. *BG.* 6.40.
172. Caes. *BCiv.* 1.44.
173. Caes. *BCiv.* 1.46.
174. Caes. *BCiv.* 1.39.
175. Caes. *BAfr.* 77.
176. Lendon, 'The Rhetoric of Combat: Greek Theory and Roman Culture in Caesar's Battle Descriptions', p. 283f.
177. *Ibid.*, pp. 284–285.
178. *Ibid.*, p. 285f.
179. *Ibid.*, p. 296f.
180. Pol. 6.24.9; Lendon, 'The Rhetoric of Combat: Greek Theory and Roman Culture in Caesar's Battle Descriptions', p. 299.
181. Lendon, 'The Rhetoric of Combat: Greek Theory and Roman Culture in Caesar's Battle Descriptions', pp. 206–8.
182. Sall. *Iug*, 52.2.
183. Lendon, 'The Rhetoric of Combat: Greek Theory and Roman Culture in Caesar's Battle Descriptions', p. 309.
184. *Ibid.*, pp. 314–16.
185. *Ibid.*, p. 325.

Chapter 8

1. T. Mommsen, *A History of Rome under the Emperors, Based on the Lecture Notes of Sebastian and Paul Hensel, 1882–6* (London, 1996), p. 92f. Cary, *A History of Rome down to the Reign of Constantine*, pp. 505–6; Smith, *Service in the Post-Marian Roman Army*, p. 70. Webster, *The Roman Imperial Army*, p. 26. Scullard, *From the Gracchi to Nero* (1976), p. 481. Keppie, *The Making of the Roman Army*, p. 260. Junkelmann, *Die Legionen des Augustus*, p. 87. Alston, *Soldier and Society in Roman Egypt: A Social History*, pp. 2–3. Andrew Wallace-Hadrill, 'Mutatas Formas: The Augustan Transformation of Roman Knowledge', in Karl Galinsky, ed., *The Cambridge Companion to the Age of Augustus* (Cambridge, 2005), p. 75. Gilliver, 'The Augustan Reform and the Structure of the Imperial Army', pp. 184–5. Edward Dąbrowa, 'The Roman Army in Syria under Augustus and Tiberius', *Limes XX Congresso internacional de estudios sobre la frontera*

romana – Xth International Congress of Roman Frontier Studies, Leon, Septiembre 2006 (2009), p. 997. Simon James, *Rome & the Sword: How Warriors and Weapons Shaped Roman History* (London, 2011), p. 125f. Adrian Keith Goldsworthy, *Augustus: First Emperor of Rome* (New Haven, 2014), pp. 348, 350. Mary Beard, *SPQR: a History of Ancient Rome* (London, 2015), p. 252. Myles Lavan, 'The Army and the Spread of Roman Citizenship', *Journal of Roman Studies*, 109 (2019), p. 28. Whately, *An Introduction to the Roman Military: From Marius (100 BCE) to Theodosius II (450 CE)*, Ch. 2, 'Augustus and the Julio-Claudians'.
2. H. Delbrück, *Geschichte der Kriegskunst im Rahmen der politischen Geschichte: die Germanen* (vol. II, Berlin, 1902), II, p. 160. von Domaszewski, *Die Rangordnung des Römischen Heeres*, p. 125.
3. Cary, *A History of Rome down to the Reign of Constantine*, pp. 505–7.
4. Scullard, *From the Gracchi to Nero* (1976), p. 251–4.
5. Keppie, *The Making of the Roman Army*, p. 146.
6. Jonathan P. Roth, *Roman Warfare* (Cambridge, 2009), p. 49.
7. Whately, *An Introduction to the Roman Military: From Marius (100 BCE) to Theodosius II (450 CE)*.
8. For example: Webster, *The Roman Imperial Army*, pp. 24–7, Lasse Zipfel, *Augustus Heeresreform. Das Heer als Grundlage des römischen Imperiums* (2015), Feugère, *Les armes des romains de la république à l'antiquité tardive*, p. 173.
9. Le Bohec, *The Imperial Roman Army*, p. 182.
10. Susan P. Mattern, *Rome and the Enemy: Imperial Strategy in the Principate* (Berkeley, 1999), pp. 126–7.
11. Junkelmann, *Die Legionen des Augustus*, pp. 99–100.
12. K.A. Raaflaub, L.J.F. Keppie and W.S. Hanson, 'The Political significance of Augustus' military reforms', *Roman Frontier Studies 1979 XII* (1980), pp. 1005–25.
13. E.A. Judge, *The Failure of Augustus: Essays on the Interpretation of a Paradox* (Newcastle, 2019), p. 80.
14. For example: Syme, *The Roman Revolution*, p. 9, A. H. M. Jones, *Augustus* (London, 1970), pp. 163–7 and David Shotter, *Augustus Caesar* (London, 1991), pp. 91–3.
15. viz.: E. Badian, '"Crisis Theories" and the Beginning of the Principate', in Johannes Heinrichs, Karl-Heinz. Schwarte and Gerhard Wirth, eds., *Romanitas-Christianitas: Untersuchungen zur Geschichte und Literatur der römischen Kaiserzeit. Johannes Straub zum 70. Geburtstag am 18. Oktober 1982 gewidmet* (Berlin, 1982), p. 19. W. K. Lacey, *Augustus and the Principate: The Evolution of the System* (Leeds, 1996). Judge, *The Failure of Augustus: Essays on the Interpretation of a Paradox*.
16. Cass. Dio 53.12.4 indicates that Augustus did not control Macedonia and 54.34 shows that he did not initially have control of Dalmatia.
17. An example of the latter problematic approach is found in Raaflaub, Keppie and Hanson, 'The Political significance of Augustus' military reforms', passim.
18. G. Webster, *Roman Army* (1956), p. 25.
19. Benjamin H. Isaac, *The Limits of Empire: The Roman Army in the East* (Oxford, 1990), p. 315.
20. Kromayer and Veith, *Heerwesen und Kriegsführung der Griechen und Römer*, p. 477. Sandra J. Bingham, 'The Praetorian Guard in the Political and Social Life of Julio-Claudian Rome', (D thesis, Vancouver, 1997), p. 12f.
21. Festus, 223.
22. Appian, *BCiv* 3.40. Suet, *Aug* 10.3.

23. Appian, *BCiv* 5.3, cf 5.59.
24. Keppie, *The Making of the Roman Army*, pp. 127, 228, plate 12c.
25. Orosius 6.19.8.
26. Bingham, *The Praetorian Guard in the Political and Social Life of Julio-Claudian Rome*, p. 25f. Suet. *Aug* 49.
27. Mommsen, *A History of Rome under the Emperors, Based on the Lecture Notes of Sebastian and Paul Hensel, 1882–6*, p. 96.
28. Keppie, *The Making of the Roman Army*, p. 151.
29. Saddington, *The Development of the Auxiliary Forces from Caesar to Vespasian*, pp. 2, 196.
30. Webster, *The Roman Imperial Army*, p. 142.
31. Pol. 6.21 describes the procedure he had seen.
32. Goldsworthy, *The Roman Army at War: 100 BC-AD 200*, p. 70f.
33. The pay rates of the members of units of *auxilia* have been difficult to determine: Cheesman, *The Auxilia of the Roman Army*, p. 36, G. Webster, *The Roman Imperial Army of the First and Second Centuries A.D.*, (Norman, 1998), 1985), p. 268. But see Phang, *Military Service. Ideologies of Discipline in the Late Republic and Early Principate*, pp. 169–70.
34. Cheesman, *The Auxilia of the Roman Army*, p. 71.
35. No known diplomas pre-date Claudius: Sara Elise Phang, *The Marriage of Roman Soldiers (13 B.C.-A.D. 235): Law and Family in the Imperial Army* (Boston, Mass., 2001), p. 53.
36. Cheesman, *The Auxilia of the Roman Army*, pp. 25, 46.
37. Veg. *Mil.* 1.8. Delbrück, *Warfare in Antiquity: History of the Art of War*, p. 161. Delbrück's reference to regulations probably refers to what are known as *Heeresdienstvorschriften*. These are known in the British and US Armies as Field Service Regulations. These documents and similar publication in other countries were used to promote improvements in the operational efficiency of European Nineteenth and Twentieth Century armies. Their impact remains considerable: Bruce Condell and David T. Zabecki, *On the German Art of War: Truppenführung* (Boulder, 2001), pp. 9–11. On the *constitutiones*, see: Jonathan Roth, *The Logistics of the Roman Army at War: (264 B.C.-A.D. 235)* (Boston (Mass.), 1999).
38. Phang, *Military Service. Ideologies of Discipline in the Late Republic and Early Principate*, pp. 3–4.
39. Parker, *The Roman Legions*, p. 288. Phang, *The Marriage of Roman Soldiers (13 B.C.-A.D. 235): Law and Family in the Imperial Army*, p. 345, fn 5.
40. *Ibid.*, p. 49.
41. *Ibid.*, p. 347f.
42. Goldsworthy, *The Roman Army at War: 100 BC-AD 200*, pp. 279–82.
43. Chester G. Starr, *The Roman Imperial Navy, 31 B.C.-A.D. 324* (Ithaca, 1941), p. 1.
44. Christian Courtois, 'Les politiques navales de l'empire romain', *Revue Historique*, 186 (1939), pp. 17–47.
45. Starr, *The Roman Imperial Navy, 31 B.C.-A.D. 324*, p. 168f.
46. Michael. Pitassi, *The Roman Navy: Ships, Men & Warfare 350 BC-AD 475* (Barnsley, 2012), pp. 26–7.
47. McArthur, 'Should Roman Soldiers be called "Professional" prior to Augustus?'.
48. *Ibid.*, p. 10.

49. Suet. *Aug* 24. The Penguin Classics translation mistakenly renders *commutavit* as reform. For *disciplina*, see Lendon, *Soldiers and Ghosts: A History of Battle in Classical Antiquity*, p. 252.
50. Augustus *RGDA* 26.1.
51. Augustus *RGDA* 26.2.
52. Augustus *RGDA* 27, 29.
53. Augustus *RGDA* 30.
54. Augustus *RGDA* 3.3.
55. Augustus *RGDA* 16.
56. Augustus *RGDA* 17.2.
57. Augustus *RGDA* 28.
58. Augustus *RGDA* Appendix 1.
59. Suet. *Aug* 20.
60. Saddington, *The Development of the Auxiliary Forces from Caesar to Vespasian*, p. 81, noting the close cooperation between the Romans and allied eastern kings.
61. Augustus simply describes what he did, not why: Aug. *RGDA* 16.1. He alludes only to those he settled in colonies: *RGDA* 3.3, 15.3. He says nothing about those he dismissed without rewards: Cass. Dio 51.3.
62. Cass. Dio 51.4.8. There may have been other attractions for Augustus in this practice either immediately after Actium or later: Phang, *The Marriage of Roman Soldiers (13 B.C.-A.D. 235): Law and Family in the Imperial Army*, p. 350f.
63. Augustus, *RGDA*. 3.3.
64. Augustus, *RGDA*. 34, 1. Alison E. Cooley, *Res Gestae Divi Augusti: Text, Translation and Commentary* (Cambridge, 2009), p. Ch. 34. Judge, *The Failure of Augustus: Essays on the Interpretation of a Paradox*, Ch. 14.
65. J.W. Rich, 'Augustus and the spolia opima', *Chiron*, 26 (1996), pp. 85–127.
66. Cass. Dio 51.24; Val. Max. 3.2.6.
67. Livy 4.20.
68. See Syme, *The Roman Revolution*, p. 308.
69. See Gallus' Philae inscription of 15 April 29: ILS 8995.
70. A.H.M. Jones, 'The Imperium of Augustus', *Journal of Roman Studies*, 41 (1951), p. 113. A.H.M. Judge, *The Failure of Augustus: Essays on the Interpretation of a Paradox*, p. 41.
71. Cass. Dio 53.25. Suet. *Aug*. 20.
72. Webster, *The Roman Imperial Army*, p. 38 fn. 1. On Augustus' wars in Spain, see R. Syme, 'The Spanish War of Augustus (26–25B.C.)', *The American Journal of Philology*, 55 (1934), pp. 293–317; and R. Syme, 'Some Notes in the Legions under Augustus', *The Journal of Roman Studies*, 23 (1933), p. 22.
73. Vell. Pat. 2.2–4.
74. Parker, *The Roman Legions*, p. 90. Florus 4.12.48.
75. Syme, 'The Spanish War of Augustus (26–25B.C.)', p. 298.
76. Vell. Pat. 2.90.1 *multo varioque Marte pacatae*. S.L. Dyson, *The Creation of the Roman Frontier* (Princeton, N.J., 1985), p. 259.
77. Cass. Dio 53.25.5.
78. Cass. Dio 53.26.1.
79. Cass. Dio 53.25.3.
80. Cass. Dio 53.25.5.

81. Cass. Dio 53.29, 54.5.
82. Cass. Dio 54.8. Vell. Pat. 2.91.1. Shotter, *Augustus Caesar*, pp. 66–7.
83. Bivar, A.D.H., 'The Political History of Iran Under the Arsacids', in Ehsan Yarshater, ed., *The Cambridge History of Iran* (Cambridge, 1983), pp. 66–7.
84. Cass. Dio 54.9.
85. Cass Dio 54.11.
86. Cass. Dio 54.11.3.
87. Suet. *Aug.* 24.
88. Cass. Dio 54.12.5.
89. Cass. Dio 54.20.
90. Vell. Pat. 2.97.1. Cass. Dio 54.20.5. Dio places the loss of the eagle in the period of the first five-year extension but Velleius places it closer to the expedition by Drusus. See also Suet. *Aug.* 23. Syme places the incident in 17: Syme, 'Some Notes in the Legions under Augustus', p. 17.
91. Cass. Dio 54.22.1
92. Cass. Dio 54.22. Vell. Pat. 2.95.1–2.
93. Cass. Dio 54.24.3.
94. CIL V, 07817. AE 1973, 00323.
95. Cass. Dio 54.25.5.
96. Pol. 6.19
97. Augustus, *RGDA* 16.1.
98. Cass. Dio 54.25.6.
99. Augustus, *RGDA* 16.1.
100. Cass. Dio 54.12.5.
101. Cass. Dio 54.28.1. The choice of Agrippa as commander, surely Augustus' best and most trusted commander, must indicate the emperor's priorities; the appointment of his stepson Tiberius to succeed Agrippa on his death confirms it: Cass. Dio 54.28.2. Vell. Pat. 2.96.1. Augustus then gave Tiberius the task of conquering Pannonia: Cass. Dio 54.35.2. Vell. Pat. 2.96.2.
102. Augustus, *RGDA* 30.
103. Scullard believes that conquest was intended: Scullard, *From the Gracchi to Nero* (1976), p. 265. Peter Wells notes the different views of Augustus' aims but, based on the archaeological evidence and ancient authors, demonstrates that Augustus intended to secure the area east of the Rhine until the Varian disaster: P.S. Wells, *The Barbarians Speak: How the Conquered Peoples of Europe Shaped the Roman Empire* (Princeton, 1999), pp. 89–92, Peter S. Wells, *The Battle that Stopped Rome: Emperor Augustus, Arminius, and the Slaughter of the Legions in the Teutoburg Forest* (New York, 2003), pp. 78–9. Colin Wells believes that Augustus aimed to conquer the territory between the Rhine and the Elbe: C. M. Wells, *The Roman Empire* (London, 1984), pp. 72–3. His views are detailed more fully in C.M. Wells, *The German policy of Augustus: An Examination of the Archaeological Evidence* (Oxford, 1972).
104. Scullard, *From the Gracchi to Nero* (1976), p. 262.
105. There is a substantial literature on the Roman frontiers with an interesting modern perspective. Dyson discusses the 'Roman frontier system' in the Late Republican and Early Imperial periods in terms of 'comparative frontier studies': Dyson, *The Creation of the Roman Frontier*, p. 3. Dyson assumes that the Romans consciously saw themselves as acculturating the native ruling class for maintenance of the frontier and

control of the interior of their provinces: *Ibid.*, p. 160. This leads him to the colonialist, anachronistic observation that 'Republican poets of Verona and Mantua have in some respects descendants on the cricket pitches of Karachi and Jamaica': *Ibid.*, p. 86. A different examination of the Roman frontiers in the light of the United States' experience comes from Drummond and Nelson. They formally acknowledge their debt to Fredrick Turner Jackson, as does Dyson, for their view of frontier as process. Drummond and Nelson believe that Augustus was forced to abandon his policy of attacking northern enemies before Rome was attacked. Instead, Augustus is presented forming 'a fixed and fortified defensive frontier in the West': S.K. Drummond and L.H. Nelson, *The Western Frontiers of Imperial Rome* (New York, 1994), p. 7. Mann argues that the views that frontiers during the principate were a product of a coherent frontier policy are erroneous: J.C. Mann, 'The Frontiers of the Principate', in H. Temporini, ed., *Aufstieg und Niedergang der romischen Welt 2, 1* (1974), p. 574. Luttwak has applied modern strategic concepts to the Romans: E.N. Luttwak, *The Grand Strategy of the Roman Empire* (Baltimore, 2016). His views, although supported aggressively by Wheeler in Wheeler, 'Methodological Limits and the Mirage of Roman Strategy: Part 1' and E.L. Wheeler, 'Methodological Limits and the Mirage of Roman Strategy: Part 2', *The Journal of Military History*, 57 (1993), pp. 215–40 and by Lacey: Lacey, 'The grand strategy of the Roman Empire' and by Ferrill: Ferrill, *Roman Imperial Grand Strategy.*, have not been accepted by others, notably Isaac in Isaac, *The Limits of Empire: the Roman Army in the East*; and by Whittaker in Whittaker, *Rome and its Frontiers: The Dynamics of Empire.*
106. Cass. Dio 54.33. Wells, *The German policy of Augustus: An Examination of the Archaeological Evidence.*
107. Austin and Rankov suggest that Augustus may have had little idea of how far in the Germany he would be able to go: Austin and Rankov, *Exploratio: Military and Political Intelligence in the Roman World from the Second Punic War to the Battle of Adrianople*, p. 127.
108. Cass. Dio 54.36.3–5, 55.1.4.
109. Cass. Dio 54.34.4.
110. Cass. Dio 54.34.5–7. Vell. Pat. 2.98.
111. Cass. Dio 54.35–6. Augustus, *RGDA* 11.3.
112. Cass. Dio 54.36.2.
113. Cass. Dio 55.6.1.
114. Cass. Dio 55.8.3.
115. Cass. Dio 55.10a.2–3.
116. Augustus, *RGDA* 17.2. Cass. Dio 55.23.1, 55.24.9, 55.25. Suet. *Aug.* 49.
117. Suet. *Aug.* 49.
118. Cass. Dio 55.23.1, 55.24.9, 55.25.
119. The existence of peaks and troughs in the demands on the *Aerarium Militare* has been noticed: P. Brennan, 'A Rome away from Rome: Veteran Colonists and Post-Augustan Roman Colonisation', in J-P Descoeudres, ed., *Proceedings of the First Congress of Classical Archaeology held in honour of Emeritus Professor A.D. Trendall, Sydney 9–14 July 1989* (Oxford, 1990), p. 496.
120. Augustus, *RGDA* 16.2.
121. G. R. Watson, *The Roman Soldier* (London, 1969), p. 147.
122. Cass. Dio 55.28.5–7. Vell. Pat. 2.105–10.

123. Sherk, while acknowledging the work of Roman surveyors, notes that there is no direct evidence that the surveyors' bronze maps, sent back to Rome, were used to prepare maps of a more general nature: Robert K Sherk and Hildegard Temporini, 'Roman geographical exploration and military maps', *Politische Geschichte (Allgemeines)*, (1974), p. 561. Janni argues that the Romans had no conception of 'horological space' as opposed to our concept of topographical space: P. Janni, *La mappa e il periplo, cartografia antica e spazio odologico* (Roma, 1984), p. 213. On the Romans' image of their world, see Mattern, *Rome and the Enemy: Imperial Strategy in the Principate*, p. Ch. 2. Arnaud argues strongly that whatever was in the Porticus Vipsania, it was of no use as a military planning tool: Pascal Arnaud, 'Marcus Vipsanius Agrippa and his Geographical Work', in Serena Bianchetti, Michele R. Cataudella and Hans-Joachim Gehrke, eds., *Brill's Companion to Ancient Geography: The Inhabited World in Greek and Roman Tradition* (Leiden, 2016), pp. 210, 216, 220–21. Rathmann argues that ancient maps were largely ignored in decision-making: Michael Rathmann, 'The *Tabula Peutingeriana* and Antique Cartography', in Bianchetti, Cataudella and Gehrke, eds., *Brill's Companion to Ancient Geography: The Inhabited World in Greek and Roman Tradition*, p. 337.
124. Cass. Dio 55.30.1. Vell. Pat. 2.110.
125. Cass. Dio 55.30.1.
126. Cass. Dio 55.30.
127. Cass. Dio 55.31.1. Vell. Pat. 2.111.1.
128. Cass. Dio 55.33.1, 55.34.4–7, 56.12.1. Vell. Pat. 2.115.1.
129. Cass. Dio 55.34.3.
130. Cass. Dio 56.12–17.
131. Cass. Dio 56.18–22. Vell. Pat. 2.118.1.
132. Tony Clunn, *The Quest for the Lost Legions: The Varusschlacht* (London, 1999).
133. Vell. Pat. 2.119.1. Suet. *Aug.* 23.
134. Suet. *Aug.* 23.
135. Suet. *Aug.* 25
136. Cass. Dio 56.23.
137. Cass. Dio 56.24.6. Vell. Pat. 2.120.1.
138. Cass. Dio 56.25. Vell. Pat. 2.121.1.
139. Augustus, *RGDA* 26.2.
140. Cass. Dio 56.28.1.
141. Cass. Dio 56.28.2–3.
142. Cass. Dio 56.28.4.
143. Cass. Dio 56.28.6.
144. Tac. *Ann.* 1.11, Dio 56.33.5. Whittaker thinks it is absurd to believe that Augustus could expect all future Roman commanders to forgo the chance of expanding the empire: Whittaker, *The Frontiers of the Roman Empire: A Social and Economic Study*, pp. 35–6.
145. Cass. Dio 56.33.2.
146. Cass. Dio 56.33.5; Tac. *Ann.* 1, 11. The nature of Augustus' advice has generated debate. Luttwak criticises Augustus for his lack of a perimeter defence: Luttwak, *The Grand Strategy of the Roman Empire*, pp. 19, 26. Isaac saw the advice to put a stop to expansion as not even true for Augustus and may only have suggested a temporary pause: Isaac, *The Limits of Empire: The Roman Army in the East*, pp. 1, 28. Whittaker

explains the advice to be consistent with Augustus practice of not occupying an area, as opposed to ruling, unless it was profitable: Whittaker, *Rome and its Frontiers: the Dynamics of Empire*, p. 41.
147. Cass. Dio 57.2ff.
148. Tac. *Ann.* 1.7.
149. Cass. Dio 57.4.1–2
150. Vell. Pat. 2.125.2. Judge, *The Failure of Augustus: Essays on the Interpretation of a Paradox*, p. 198.
151. Tac. *Ann.* 1.17.
152. J. B. Campbell, *The Emperor and the Roman Army, 31 BC–AD 235* (Oxford, 1984), p. 28.
153. Hans-Christian Schneider, *Das Problem der Veteranenversorgung in der späteren römischen Republik* (Bonn, 1977), p. 239.
154. Tac. *Ann.* 1,17.
155. Tac. *Ann.*1.34.
156. Tac. *Ann.* 1.35.
157. Tac. *Ann.* 1.36.
158. Tac. *Ann.* 1.36–7.
159. Tac. *Ann.* 4.4.

Conclusion
1. P. Erdkamp, *Hunger and the Sword: Warfare and Food Supply in Roman Republican Wars (264–30 B.C.)* (Amsterdam, 1998), p. 1.
2. *Ibid.*, p. 3.

Bibliography

Abbott, F.F., *A Short History of Rome* (New York, 1906).
Adam, A-M., 'Emprunts et échanges de certains types d'armement entre l'Italie et le monde non-méditerranéen aux Ve et IVe siècles avant J-C, in A-M. Adam, and A. Rouveret, ed., *Guerre et Sociétés en Italie aux Ve et IVe Siècles avant J-C* (Paris, 1986), pp. 19–28.
Adam, A-M., and A. Rouveret, ed., *Guerre et sociétés en italie aux Ve et Ive siècles ave J-C* (Paris, 1986).
Adams, Colin, 'War and Society', in Philip Sabin, Hans Van Wees, and Michael Whitby, eds., *The Cambridge History of Greek and Roman Warfare: Volume II Rome from the late Republic to the late Empire* (Cambridge, 2007), pp. 198–232.
Adcock, F. E., *The Roman Art of War under the Republic* (Cambridge, Mass, 1940).
Aelian, *The Tactics of Aelian or On the Military Arrangements of the Greeks* (Barnsley, 2012).
G. Alföldy, 'The crisis of the third century as seen by contemporaries', *Greek, Roman, and Byzantine Studies*, 15 (1974), pp. 89–111.
Allmand, C. T., *The De Re Militari of Vegetius: The Reception, Transmission and Legacy of a Roman Text in the Middle Ages* (Cambridge, 2011).
Alston, R., *Soldier and Society in Roman Egypt: A Social History* (London; New York, 1995).
———, *Aspects of Roman History, AD 14–117* (London, 1998).
———, 'The Military and Politics', in Philip Sabin, Hans Van Wees, and Michael Whitby, eds., *The Cambridge History of Greek and Roman Warfare: Volume II Rome from the late Republic to the late Empire* (Cambridge, 2007), pp. 176–197.
Armstrong, Jeremy, 'Organised Chaos: *Manipuli, Socii* and the Roman Army C.300', in Jeremy Armstrong, and Michael P. Fronda, eds., *Romans at War: Soldiers, Citizens, and Society in the Roman Republic* (London, 2020), pp. 76–98.
Armstrong, Jeremy, and Michael P. Fronda, 'Writing About Romans at War', in Jeremy Armstrong, and Michael P. Fronda, eds., *Romans at War: Soldiers, Citizens, and Society in the Roman Republic* (London, 2020), pp.1–16.
Arnaud, Pascal, 'Marcus Vipsanius Agrippa and His Geographical Work', in Serena Bianchetti, Michele R. Cataudella, and Hans-Joachim Gehrke, eds., *Brill's Companion to Ancient Geography: The Inhabited World in Greek and Roman Tradition* (Leiden, 2016), pp. 205–22.
Arnold, T., *The History of Rome* (vol. I, New York, 1846).
———, *History of Rome*, Volume II (vol. II, London, 1848).
A. E. Astin, F. W. Walbank, M. W. Frederiksen, and R. M. Ogilvie, eds., *The Cambridge Ancient History Volume 8: The Cambridge Ancient History: Rome and the Mediterranean to 133 BC* (Cambridge, 1989).
Austin, N. J. E, and N. B Rankov, *Exploratio: Military and Political Intelligence in the Roman World from the Second Punic War to the Battle of Adrianople* (London, 1995).
Badian, E., *Foreign Clientelae* (Oxford, 1958).

———, *Publicans and Sinners* (Oxford, 1972).
———, '"Crisis Theories" and the Beginning of the Principate', in Johannes Heinrichs, Karl-Heinz. Schwarte, and Gerhard Wirth, eds., *Romanitas-Christianitas: Untersuchungen zur Geschichte und Literatur der römischen Kaiserzeit. Johannes Straub zum 70. Geburtstag am 18. Oktober 1982 gewidmet* (Berlin, 1982), pp. 18–41.
Beard, Mary, *SPQR: A History of Ancient Rome* (London, 2015).
M.J.V. Bell, 'Tactical Reform in the Roman Republican Army', *Historia*, 14 (1965), pp. 404–422.
Berenhorst, Georg Heinrich von, *Betrachtungen über die Kriegskunst, über ihre Fortschritte, ihre Widersprüche und ihre Zuverläßigkeit. Auch für Layen verständlich, wenn sie nur Geschichte wissen* (Leipzig, 1797).
Berneck, Karl Gustav von, *Geschichte der Kriegskunst für Militairakademien und Offiziere aller Grade* (Berlin, 1867).
Bingham, Sandra J., 'The Praetorian Guard in the Political and Social Life of Julio-Claudian Rome', (Ph.D thesis, Vancouver, 1997).
Eric Birley, 'Hadrian's Wall', *The Antiquaries Journal*, XI (1931), p. 62.
Bishop, M.C., and J.C.N. Coulston, *Roman Military Equipment from the Punic Wars to the Fall of Rome* (Oxford, 2006).
Bivar, A.D.H., 'The Political History of Iran Under the Arsacids', in Ehsan Yarshater, ed., *The Cambridge History of Iran* (Cambridge, 1983), pp. 21–99.
Bowman, Alan, Averil Cameron, and Peter Garnsey, eds., *The Cambridge Ancient History Volume 12: Crisis of Empire, AD 193–337* (Cambridge, 2005).
Bowman, Alan K., Peter Garnsey, and Dominic Rathbone, eds., *The Cambridge Ancient History Volume 11: High Empire, A.D. 70–192* (Cambridge, 2000).
Bradford, Jeffery A., *Proconsuls and CinCs from the Roman Republic to the Republic of the United States of America: Lessons for the Pax Americana* (Monograph, School of Advanced Military Studies United States Army Command and General Staff College Fort Leavenworth, Kansas, 2001).
Bradley, Guy Jolyon, *Early Rome to 290 BC: The Beginnings of the City and the Rise of the Republic* (Edinburgh, 2020).
Breeze, David J., *The Roman Army* (London, 2016).
Brennan, P., 'A Rome Away from Rome: Veteran Colonists and Post-Augustan Roman Colonisation', in J-P Descoeudres, ed., *Proceedings of the First Congress of Classical Archaeology held in honour of Emeritus Professor A.D. Trendall, Sydney 9–14 July 1989* (Oxford, 1990), pp. 491–502.
Brennan, P., T.W. Hillard, R.A. Kearsley, C.V.E. Nixon, and A.M. Nobbs, 'Divide and Fall: The Separation of Legionary Cavalry and the Fragmentation of the Roman Empire', *Ancient History in a Modern University* (2, 1998), pp. 238–44.
Brice, Lee L., '*SPQR* SNAFU: Indiscipline and Internal Conflict in the Late Republic', in Jeremy Armstrong, and Michael P. Fronda, eds., *Romans at War: Soldiers, Citizens, and Society in the Roman Republic* (London, 2020), pp. 232–246.
Briquel, D., 'La tradition sur l'emprunt d'armes samite par Rome', in A.-M. Adam, and A. Rouveret, ed., *Guerre et sociétés en Italie aux Ve et IVe Siècles avant J-C* (Paris, 1986), pp. 65–90.
Briquel, Dominique., *Mythe et révolution : a fabrication d'un récit : la naissance de la république à Rome* (Bruxelles, 2007).
Brizzi, Giovanni, *Le guerrier de l'antiquité classique : de l'hoplite au légionnaire* (Monaco, 2004).

Brunsson, Nils, *Reform as Routine: Organizational Change and Stability in the Modern World* (Oxford, 2009).
P. A. Brunt, 'The army and the land in the Roman revolution', *The Journal of Roman Studies*, LII (1962), pp. 69–86.
Brunt, P.A., *Italian Manpower* (Oxford, 1971).
Bucholz, A., *Hans Delbrück and the German Military Establishment: War Images in Conflict* (Iowa City, 1985).
Burckhardt, Leonhard, *Militärgeschichte der Antike* (2008).
Burns, M.T., 'The Homogenisation of Military Equipment under the Roman Republic', *Digressus Supplement 1* "Romanisation?" (2003), pp. 60–85.
Bury, J. B., *The Idea of Progress: An Inquiry into Its Origin and Growth* (London, 1920).
Cadiou, F., *L'armée imaginaire : les soldats prolétaires dans les légions romaines au dernier siècle de la république* (Paris, 2018).
Cadiou, François, 'Les guerres en *Hispania* et l'émergence de la cohorte légionnaire dans l'armée romaine sous la république: une révision critique', *Gladius*, XXI (2001), pp. 167–82.
Caforio, Giuseppe, ed., *Handbook of the Sociology of the Military* (New York, 2006).
Cagnat, René, *L'armée romaine d'Afrique et l'occupation militaire de l'Afrique sous les empereurs* (Paris, 1913).
Cagniart, Pierre, 'The Late Republican Army (146–30BC)', in P. Erdkamp, ed., *Companion to the Roman Army* (Oxford, 2007), pp. 80–96.
Cameron, Averil, and Peter Garnsey, *The Cambridge Ancient History Volume 13: The Late Empire, AD 337–425* (Oxford, 1998).
Campbell, Duncan B., 'Arming the Romans for Battle', in Brian Campbell, and Lawrence A. Tritle, eds., *The Oxford Handbook of Warfare in the Classical World* (Oxford, 2013), pp. 2–14.
Campbell, J. B., *The Emperor and the Roman Army, 31 BC–AD 235* (Oxford, 1984).
———, 'Teach yourself how to be a general', *Journal of Roman Studies*, 77 (1987), pp. 13–28.
Capes, W.W., *Roman History: The Early Empire* (London, 1897).
Carley, Lionel Kenneth, 'The Anglo-Norman Vegetius: A Thirteenth Century Translation of the "de Re Militari"', (Ph.D. Thesis, Nottingham, 1962).
Carrion-Nisas, Colonel Marie Henri François, *Essai sur l'histoire générale de l'art militaire, de son origine, des ses progrès et des ses révolutions, depuis la première formation des sociétés européennes jusqu'à nos jours, orné de quatorze places* (vol. 2, Paris, 1824).
Cary, M., *A History of Rome Down to the Reign of Constantine* (London, 1938).
Dio, Cassius, trans. E. Cary, *Dio's Roman History, in Nine Volumes* (Cambridge, Mass., 1968).
Chaplin, Jane D., *Livy's Exemplary History* (Oxford, 2000).
Chariotis, Angelos, 'The Impact of War on the Economy of Hellenistic *Poleis*: Demand Creation, Short-Term Influences and Long-Term Impacts', in Zosia H. Archibald, John K. Davies, and Vincent. Gabrielsen, eds., *The Economies of Hellenistic Societies, Third to First Centuries BC* (Oxford, 2011), pp. 122–141.
Cheesman, G.L., *The Auxilia of the Roman Army* (Oxford, 1914).
Childs, John, *Armies and Warfare in Europe, 1648–1789* (Manchester, 1982).
Cichorius, Conrad, *Die Reliefs der Traianssäule* (Vols 1 and 2, Berlin, 1896).
Clark, Christopher M., *Iron Kingdom: The Rise and Downfall of Prussia, 1600–1947* (London, 2007).

Clark, Edward Dale, 'Roman legionary forces in Sicily during the Second Punic War: the number of legions stationed on the island from 214 to 210 B.C.', *The Ancient History Bulletin*, 8 (1994), pp. 133–140.
Clunn, Tony, *The Quest for the Lost Legions: The Varusschlacht* (London, 1999).
Cobban, A., *A History of Modern France* (vol. 3, Hammondsworth, 1965).
Condell, Bruce, and David T. Zabecki, *On the German Art of War: Truppenführung* (Boulder, 2001).
Cooley, Alison E., *Res Gestae Divi Augusti: Text, Translation and Commentary* (Cambridge, 2009).
Cornell, Timothy J., *The Beginnings of Rome: Italy and Rome from the Bronze Age to the Punic Wars (C 1000–264 BC)* (London, 1995).
Cosme, Pierre, *L'armée romaine : VIIIe s. av. J.-C.-Ve s. ap. J.-C* (Paris, 2012).
Couissin, P., *Les armes romaines* (Paris, 1926).
Courtois, Christian, 'Les politiques navales de l'empire romain', *Revue Historique*, 186 (1939), pp. 17–47.
Cowan, R., 'Etruscan and Gallic Pila: origins of the heavy javelin', *Ancient Warfare*, XII-6 (2019), pp. 18–21.
Cowan, Ross, and Adam Hook, *Roman Battle Tactics, 109 BC–AD 313* (Oxford, 2007).
Crawford, M., *The Roman Republic* (Cambridge, Mass., 1992).
Crawford, Michael H., *Coinage and Money Under the Roman Republic: Italy and the Mediterranean Economy* (London, 1985).
Crawford, Michael H., and J. D. Cloud, *Roman Statutes* (London, 1996).
Crook, J. A., Andrew Lintott, and Elizabeth Rawson, *The Cambridge Ancient History Volume IX: The Last Age of the Roman Republic, 146–43 BC* (Cambridge, 1994).
D'Amato, Raffaele., and Graham. Sumner, *Arms and Armour of the Imperial Roman Soldier: From Marius to Commodus, 112 BC–AD 192* (London, 2009).
Dąbrowa, Edward, 'The Roman Army in Syria under Augustus and Tiberius', *Limes XX Congresso internacional de estudios sobre la frontera romana – Xth International Congress of Roman Frontier Studies, Leon, Septiembre 2006* (Madrid, 2009), pp. 997–1005.
Dague, Everett Thomas, *Napoleon and the First Empire's Ministries of War and Military Administration: The Construction of a Military Bureaucracy* (Lewiston, N.Y., 2006).
Dawson, Doyne., *The Origins of Western Warfare: Militarism and Morality in the Ancient World* (Boulder, Colo., 1996).
de Blois, Lukas, 'Army and General in the Late Roman Republic', in P. Erdkamp, ed., *A Companion to the Roman Army* (Oxford, 2007), pp. 164–180.
de Cazanove, Olivier, 'Pratiques et rites de la guerre en Italie, entre romains et samnites: le passage sous le joug, la légion de lin samnite', in J-Chr. Couvenhes, S. Crouzet, and S. Péré-Noguès, eds., *Pratiques et identités culturelles des armées hellénistiques du monde méditerranéen : Hellenistic Warfare 3* (Bordeaux, 2011), pp. 357–401.
de Folard, Jean Charles Chevalier de., *Mémoires pour servir à l'histoire de monsieur le chevalier de Folard* (Ratisbonne, 1753).
de Guibert, Jacques-Antoine-Hippolyte Comte, *Oeuvres militaries de Guibert* (vol. 1, Paris, 1803).
de Jomini, Baron A.H., *Traité des grandes opérations militaires, contenant l'histoire critique des campagnes de Frédéric ii, comparées à celles de l'empereur Napoléon : avec un recueil des principes généraux de l'art de la guerre* (vol. I, II and III, Paris, 1811).

de Jomini, Baron, *Précis de l'art de la guerre, ou nouveau tableau analytique des principales combinations de la stratégie, de la grande tactique et de la politique militaire* (vol. Partie I, Paris, 1837).
——, *Précis de l'art de la guerre, ou nouveau tableau analytique des principales combinations de la stratégie, de la grande tactique et de la politique militaire* (vol. Partie II, Paris, 1838).
——, *Précis de l'art de la guerre, or nouveau traité analytique des principales combinations de la stratégie, de la grande tactique et de la politique militaire* (vol. II, Bruxelles, 1840).
——, trans. Mendell, Capt. G.H. and Craighill, Lieut. W.P., *The Art of War by Baron de Jomini* (Philadelphia, 1862).
——, trans. Maj. W.O. Winship and Lieut. E.E. McLean, *Summary of the Art of War Or, a New Analytical Compend of the Principal Combinations of Strategy, of Grand Tactics and of Military Policy* (New York, 1854).
de La Barre Duparcq, Nicolas Édouard, *Éléments d'art et d'histoire militaires* (Paris, 1858).
——, trans. Brig-Gen. George W. Cullum, E., *Duparcq's Military Art and History* (New York, 1983).
de la Chauvelays, Jules, *L'art militaire chez les romains. Nouvelles observations critiques. Pour faire suite à celles du chevalier Folard et du colonel Guiscard avec une lettre du général Davout, duc d'Auerstedt* (Paris, 1884).
De Saxe, M., *Mes rêveries* (vol. 1, Amsterdam, 1757).
Delbrück, H., *Geschichte der Kriegskunst im Rahmen der politischen Geschichte: die Germanen* (vol. II, Berlin, 1902).
——, *Geschichte der Kriegskunst im Rahmen der politischen Geschichte: das Altertum* (vol. I, Berlin, 1920).
——, *Warfare in Antiquity: History of the Art of War* (vol. 1, Lincoln, 1990).
Dennis, George T., *The Taktika of Leo VI* (Washington, 2014).
Derow, Peter, Andrew Erskine, and Josephine Crawley Quinn, eds., *Rome, Polybius, and the East* (Oxford, 2015).
Deutsche Akademie Der Wissenschaften, *Corpus Inscriptionum Latinarum* (Berlin, 1893).
Dixon, K.R., and P. Southern, *The Roman Cavalry* (London, 1992).
Dobson, Michael J., *The Army of the Roman Republic: The Second Century BC, Polybius and the Camps at Numantia, Spain* (Oxford, 2008).
Dodds, E.R., *The Ancient Concept of Progress* (Oxford, 1973).
Dodge, T.A., *Caesar: A History of the Art of War Among the Romans Down to the End of the Roman Empire* (Boston, New York, 1892).
Drogula, Fred K., *Commanders and Command in the Roman Republic and Early Empire* (Chapel Hill, 2015).
Drummond, S.K., and L.H. Nelson, *The Western Frontiers of Imperial Rome* (New York, 1994).
Dubuisson, M., *Le latin de Polybe : les implications historiques d'un cas de bilinguisme* (Paris, 1985).
Duffy, C., *The Military Experience in the Age of Reason* (London, 1987).
Du Picq, Col. A., *Études sur le combat* (Paris, 1880).
——, trans Col. John N. Cotton and Maj. Robert C. Cotton, *Battle Studies Ancient and Modern* (New York, 1921).
Dyson, S.L., *The Creation of the Roman Frontier* (Princeton, N.J., 1985).
Echeverría Rey, Fernando, 'Weapons, Technological Determinism and Ancient Warfare', in Garrett G. Fagan, and Matthew Trundle, eds., *New Perspectives on Ancient Warfare* (Leiden, 2010), pp. 21–56.

Eck, Werner, *Augustus und seine Zeit* (München, 2014).
Eckstein, A.M., *Mediterranean Anarchy, Interstate War, and the Rise of Rome* (Berkeley, 2009).
Eckstein, Arthur M., *Senate and General: Individual Decision Making and Roman Foreign Relations 264–194 B.C.* (Berkeley; London, 1987).
Eisenstein, Elizabeth L., *The Printing Revolution in Early Modern Europe* (Cambridge, 2013).
Elliot, Simon, *Romans at War: The Roman Military in the Republic and Empire* (Oxford, 2020).
Enenkel, K. A. E., Koen Ottenheym, and Alexander C. Thomson, *Ambitious Antiquities, Famous Forebears: Constructions of a Glorious Past in the Early Modern Netherlands and in Europe* (Leiden, 2019).
Erdkamp, P., *Hunger and the Sword: Warfare and Food Supply in Roman Republican Wars (264–30 B.C.)* (Amsterdam, 1998).
Erdkamp, Paul, 'The Corn Supply of the Roman Armies during the Third and Second Centuries B.C', *Historia: Zeitschrift für alte Geschichte*, 44 (1995), pp. 168–191.
———, Army and Society', in Nathan Rosenstein, and Robert Morstein-Marx, eds., *A Companion to the Roman Republic* (Malden, 2006), pp. 278–296.
———, 'Polybius and Livy on the Allies in the Roman Army', in Lukas De Blois, Elio Lo Cascio, Olivier Hekster, and Gerda De Kleijn, eds., *Impact of the Roman Army (200 BC–AD 476): Economic, Social, Political, Religious, and Cultural Aspects: Proceedings of the Sixth Workshop of the International Network Impact of Empire (Roman Empire 200 BC–AD 476), Capri, March 29-April 2, 2005* (Leiden; Boston (Mass.), 2007), pp. 47–74.
Erskine, Andrew, 'How to Rule the World: Polybius Book 6 Reconsidered', in Bruce John Gibson, and Thomas Harrison, eds., *Polybius and His World: Essays in Memory of F. W. Walbank* (Oxford, 2013), pp. 231–245.
Evetts, Julia, 'Explaining the Construction of Professionalism in the Military: History, Concepts and Theories', *Ophrys: revue française de sociologie*, 44 (2003), pp. 759–766.
Farnum, J.H., *The Positioning of the Roman Imperial Legions* (Oxford, 2005).
Feaver, Peter, *Armed Servants: Agency, Oversight, and Civil-Military Relations* (Cambridge, Mass.; London, 2003).
Feldheer, A., *Spectacle and Society in Livy's History* (Berkeley, 1998).
Ferguson, Adam, *The History of the Progress and Termination of the Roman Republic* (vol. III, London, 1783).
Ferrill, Arthur., *Roman Imperial Grand Strategy* (Lanham, 1991).
Feugère, M., *Les armes des romains de la république à l'antiquité tardive* (Paris, 2002).
Finer, S. E., *The Man on Horseback: The Role of the Military in Politics* (New Brunswick, 2002).
Firges, Pascal, Johna Lange, Thomas Maissen, Sebastian Meurer, Susan Richter, Gregor Stiebert, Lina Weber, and Christine Zabel, 'Languages of Reform in the European Enlightenment', in Susan Richter, Thomas Maissen, and Manuela Albertone, eds., *Languages of Reform in the Eighteenth Century: When Europe Lost Its Fear of Change* (New York, 2020), pp. 1–26.
Flynn, George Q., *Conscription and Democracy: The Draft in France, Great Britain, and the United States* (London, 2002).
Forsythe, Gary., *A Critical History of Early Rome* (Berkeley, 2005).
———, 'The Army and the Centuriate Organisation in Early Rome', in P. Erdkamp, ed., *A Companion to the Roman Army (Oxford, 2007)*, pp. 24–42.
Foss, C., *Roman Historical Coins* (London, 1990).
François, Hermann von, *Tannenberg: Das Cannae des Weltkrieges in Wort und Bild* (Berlin, 1926).

Frank, Tenney, *A History of Rome* (New York, 1923).
Franzoni, Claudio., *Habitus atque habitudo militis: monumenti funerari di militari nella cisalpina romana* (Roma, 1987).
Fröhlich, Franz, *Das Kriegswesen Cäsars* (vol. 1, Zürich, 1890).
Gabba, E., *Republican Rome, the Army and the Allies* (Oxford, 1976).
Garlan, Yvon, *War in the Ancient World: A Social History* (London, 1975).
Garnsey, P., and R. Saller, *The Roman Empire: Economy, Society, Culture* (London, 1987).
Gat, Azar, *The Origins of Military Thought from the Enlightenment to Clausewitz* (Oxford, 1989).
———, *War in Human Civilization* (Oxford, 2006).
Gat, Azar, and Alexander Yakobson, *Nations: The Long History and Deep Roots of Political Ethnicity and Nationalism* (Cambridge, 2013).
Gauthier, François, 'The Transformation of the Roman Army in the Last Decades of the Roman Republic', in Jeremy Armstrong, and Michael P. Fronda, eds., *Romans at War: Soldiers, Citizens, and Society in the Roman Republic* (London, 2020), pp. 283–96.
Gibbon, Edward., *The History of the Decline and Fall of the Roman Empire, Volume the Second* (London, 1781).
———, *The History of the Decline and Fall of the Roman Empire, Volume the Third* (London, 1781).
———, *The History of the Decline and Fall of the Roman Empire, Volume the Fourth* (London, 1788).
———, *The History of the Decline and Fall of the Roman Empire, Volume the Fifth* (London, 1788).
———, *The History of the Decline and Fall of the Roman Empire, Volume the Sixth* (London, 1788).
———, *The Decline and Fall of the Roman Empire* (London, 1995).
Gibbon, Edward, and Georges Alfred Bonnard, *Memoirs of my Life* (London, 1966).
Gilliver, C.M., 'Display in Roman Warfare: The Appearance of Armies and Individuals in the Battlefield', *War in HIstory*, 14 (2007), pp. 1–21.
Gilliver, C.M., *The Roman Art of War* (Stroud, 1999).
Gilliver, Catherine M., 'The Roman Army and the Morality of War', in A.B. Lloyd, ed., *Battle in Antiquity* (London, 1996), Ch. 6.
Gilliver, K., 'The Augustan Reform and the Structure of the Imperial Army', in P. Erdkamp, ed., A *Companion to the Roman Army* (Oxford, 2007), pp. 183–201.
Goldsmith, O., Goldsmith's Roman History: *Abridged by Himself, for the Use of Schools* (Poughkeepsie, 1816).
Goldsworthy, A, 'Instinctive Genius. The Depiction of Caesar the General', in A. Powell, and K. Welch, eds., *Julius Caesar as Artful Reporter* (London, 1998), pp. 193–220.
———, *Roman Warfare* (London, 2000).
———, *The Punic Wars* (London, 2000).
———, *The Roman Army at War: 100 BC–AD 200* (Oxford, 1996).
———, *The Complete Roman Army* (London, 2003).
———, *Augustus: First Emperor of Rome* (New Haven, 2014).
Gray, David R., 'New Age Military Progressives: U.S. Army Officer Professionalism in the Information Age', (US Army War College, Carlisle Barracks, Pennsylvania, 2001).
Gruen, E., and K. Galinsky, 'Augustus and the Making of the Principate', *The Cambridge Companion to the Age of Augustus* (Cambridge, 2005), pp. 33–54.

Gruen, Erich. S., *The Last Generation of the Roman Republic* (Berkeley, 1995).

———, 'Polybius and Josephus on Rome', in F. W. Walbank, Bruce Gibson, and Thomas Harrison, eds., *Polybius and His World: Essays in Memory of F.W. Walbank* (2013), pp. 255–66.

Guischardt, Charles, *Mémoires militaires sur les grecs et les romains, où l'on a fidélement retabli sur le texte de Polybe et des tacticiens grecs et latins* (vol. 1, La Haie, 1758).

Guischardt, Karl Gottlieb, *Principes de l'art militaire : extraits des meilleurs ouvrages des anciens* (vol. 1, Berlin, 1763).

Haldon, John F., *Warfare, State and Society in the Byzantine World 565–1204* (London, 1999).

Hanson, Victor Davis, *Hoplites: The Classical Greek Experience of Battle* (London, 1991).

———, *The Western Way of War: Infantry Battle in Classical Greece* (New York, 1989).

Harkness, A., *The Military System of the Romans* (New York, 1887).

Harmand, J., *L'armée et le soldat à Rome de 107 à 50 avant notre ère* (Paris, 1967).

Harnack, Adolf von, *Militia Christi: Die Christliche Religion und der Soldatenstand in den ersten drei Jahrhunderten* (Tübingen, 1905).

Harris, W.V., *War and Imperialism in Republican Rome, 327–70 B.C.* (Oxford, 1979).

Haverfield, F., 'Obituary: Leonard Cheesman', *Journal of Roman Studies*, 5 (1915), pp. 147–8.

Helgeland, John, 'Christians and the Roman army, A.D. 173–337', *Church History: Studies in Christianity and Culture*, XLIII (1974), pp. 149–163.

Helm, Marian, 'Poor Man's War – Rich Man's Fight. Military Integration in Republican Rome', in Jeremy Armstrong, and Michael P. Fronda, eds., *Romans at War: Soldiers, Citizens, and Society in the Roman Republic* (London, 2020), pp. 99–115.

Hildinger, Erik, *Swords Against the Senate: The Rise of the Roman Army and the Fall of the Republic* (Cambridge (Mass.), 2002).

Hirschi, Caspar, *The Origins of Nationalism: An Alternative History from Ancient Rome to Early Modern Germany* (Cambridge, 2012).

Hobsbawm, E. J., *Industry and Empire: From 1750 to the Present Day* (Harmondsworth, 1969).

Hooke, Nathaniel, *The Roman History from the Building of Rome to the Ruin of the Commonwealth* (vol. II, Dublin, 1759).

———, The Roman History from the Building of Rome to the Ruin of the Commonwealth (vol. III, Dublin, 1759).

Horvat, Jana, 'The Hoard of Roman Republican Weapons from Grad near Šmihel', *Arheološki Vestnik*, 53 (2002), pp. 117–92.

Hoyos, D., 'The Age of Overseas Expansion (264–146 B.C.)', in P. Erdkamp, ed., *A Companion to the Roman Army* (Oxford, 2007), pp.63–79.

Hughes, B.P., *Firepower* (London, 1974).

Hull, Isabel V., *Absolute Destruction: Military Culture and the Practices of War in Imperial Germany* (Ithaca, N.Y.; London, 2005).

Huntington, Samuel P., *The Soldier and the State: The Theory and Politics of Civil-Military Relations* (Cambridge, 1957).

Iggers, Georg G., Q. Edward Wang, and Supriya Mukherjee, *A Global History of Modern Historiography* (London, 2008).

Innes, Joanna, '"Reform" in English Public Life: The Fortunes of a Word', in Arthur Burns, and Joanna Innes, eds., *Rethinking the Age of Reform* (Cambridge, 2003), pp. 71–97.

Isaac, Benjamin H., *The Limits of Empire: The Roman Army in the East* (Oxford, 1990).

Jähns, Max, *Handbuch einer Geschichte des Kriegswesens von der Urzeit bis zur Renaissance: Technischer Theil: Bewaffnung, Kampfweise, Befestigung, Belagerung, Seewesen* (Leipzig, 1880).
James, S., 'Writing the Legions: The Development and Future of Roman Military Studies in Britain', *Archaeology Journal*, 159 (2002), pp. 1–58.
James, Simon, *Rome & the Sword: How Warriors and Weapons Shaped Roman History* (London, 2011).
Jomini, Baron, *Treatise on Grand Military Operations or a Critical and Military History of the Wars of Frederick the Great as Contrasted with the Modem System* (New York, 1865).
Jones, A.H.M., *Augustus* (London, 1970).
Jones, A.H.M., 'The Imperium of Augustus', *Journal of Roman Studies*, 41 (1951), pp. 112–19.
Judge, E.A., 'Second Thoughts on Augustus', *Ancient Society*, 27 (1997), pp. 43–75.
———, *The Failure of Augustus: Essays on the Interpretation of a Paradox* (Newcastle, 2019).
Judson, H.P., *Caesar's Army; a Study of the Military Art of the Romans in the Last Days of the Republic* (New York, 1888).
Junkelmann, M., *Die Legionen des Augustus* (Mainz am Rhein, 1986).
Kameka, Eugene, 'Political Nationalism - the Evolution of the Idea', in Eugene Kamenka, ed., *Nationalism: The Nature of an Idea* (Canberra, 1973), pp. 2–21.
Keaveney, Arthur, *The Army in the Roman Revolution* (London, 2007).
Keegan, J., *A History of Warfare* (London, 1993).
Kelly, C., *Ruling the Later Roman Empire* (Cambridge, Mass., 2004).
Kent, Patrick, 'Reconsidering «socii» in Roman Armies Before the Punic Wars', in Saskia T. Roselaar, ed., *Processes of Integration and Identity Formation in the Roman Republic* (Leiden, 2012), pp. 71–83.
Keppie, L., 'A centurion of legio Martia at Padova?', *Journal of Roman Military Equipment Studies*, 2 pp. 115–121.
Keppie, Lawrence, *The Making of the Roman Army* (London, 1984).
———, 'The Army and the Navy', in Alan K. Bowmans, Edward Champlin, and Andrew Lintott, eds., *The Cambridge Ancient History: Volume X The Augustan Empire, 43 B.C.- A.D. 69* (Cambridge, 1996), pp. 1056–1058.
Kertész, I., 'The Roman Cohort Tactics - Problems of Development', *Oikumene*, 1 (1976), pp. 89–97.
Klejnowski, Grzegorz, 'Hasta Velitaris: the first edge of the Roman army', *Res Militaris Studia nad wojskowością antyczną*, II (2015),
Koon, Sam, *Infantry Combat in Livy's Battle Narratives* (Oxford, 2010).
Kromayer, J., and G. Veith, *Heerwesen und Kriegsführung der Griechen und Römer* (München, 1928).
Kubitschek, W., 'Legio. Republikanische Zeit', *Paulys Realencyclopädie der classischen Altertumswissenschaft*, XII, 1 (1924), col. 1186–1210.
Lacey, James, 'The Grand Strategy of the Roman Empire', in Richard Hart Sinnreich, and Williamson Murray, eds., *Successful Strategies: Triumphing in War and Peace from Antiquity to the Present* (Cambridge, 2014), pp. 38–64.
Lacey, W. K., *Augustus and the Principate: The Evolution of the System* (Leeds, Great Britain, 1996).
Ladner, G.B., *The Idea of Reform, Its Impact on Christian Thought and Action in the Age of the Fathers* (Cambridge, Mass., 1959).

Lamarre, Clovis, *De la milice romaine depuis la fondation de Rome jusqu'à Constantin* (Paris, 1863).
Last, H., 'Gaius Gracchus', in S.A. Cook, F.E. Adcock, and M.P. Charlesworth, eds., *The Cambridge Ancient History Volume IX: the Roman Republic* (London, 1932), pp. 40–101.
———, 'Wars of the Age of Marius, in S.A. Cook, F.E. Adcock, and M.P. Charlesworth, eds., *The Cambridge Ancient History Volume IX: the Roman Republic* (London, 1932), pp. 102–57.
Lavan, Myles, 'The Army and the Spread of Roman Citizenship', *Journal of Roman Studies*, 109 (2019), pp. 27–69.
Le Bohec Y., *The Imperial Roman Army* (London, 1994).
———, 'Roman Wars and Armies in Livy, in Bernard Mineo, ed., *A Companion to Livy* (Chichester, 2015), pp. 114–124.
Le Bohec, Yann, *L'armée romaine sous le bas-empire* (Paris, 2006).
———, *Histoire des guerres romaines milieu du VIIIe siècle avant J.-C.-410 après J.-C.* (Paris, 2017).
———, ed., *The Encyclopaedia of the Roman Army* (vol. 1, Chichester, 2015).
Leighton R.F., *A History of Rome* (New York, 1889).
Lendon, J. E., *Soldiers and Ghosts: A History of Battle in Classical Antiquity* (New Haven, 2005).
Lendon, J. E., 'The Rhetoric of Combat: Greek Theory and Roman Culture in Caesar's Battle Descriptions', *Classical Antiquity*, 18 (1999), pp. 273–329.
Lewis, C.T., C. Short, E.A. Andrews, and W. Freund, *A Latin Dictionary: Founded on Andrews' Edition of Freund's Latin Dictionary* (Oxford, 1945).
Lewis, G.C., *An Inquiry into the Credibility of the Early Roman History* (vol. II, London, 1855).
Liddell, Henry G., *A History of Rome from the Earliest Times to the Establishment of the Empire* (vol. I, London, 1855).
Lindenschmit, Ludwig, *Tracht und Bewaffnung des römischen Heeres während der Kaiserzeit* (Braunschweig, 1882).
Lipowsky, Felix Joseph, *Des Flavius Vegetius Renatus fünf Bücher über die Kriegswissenschaft und Kriegskunst der Römer* (Sulzbach, 1827).
Lipsius, Justus, *Iusti Lipsi de Militia Romana Libri Quinque, Commentarius ad Polybium* (Antveria, 1598).
Livy, trans. B.O. Foster, *Livy with an English Translation in Fourteen Volumes* (Cambridge, Mass., 1982).
Löhr Carl Adolf, *Ueber die Taktik und das Kriegswesen der Griechen und Römer* (Kempten, 1825).
Luce, T.J., *Livy: The Composition of his History* (Princeton, 1977).
Luttwak, E.N., *The Grand Strategy of the Roman Empire* (Baltimore, 1976).
Lynn, John A., *Battle: A History of Combat and Culture* (New York, 2003).
Lyotard, Jean-François, *La condition postmoderne : rapport sur le savoir* (Paris, 1979).
Machiavelli, N., *The Art of War* (New York, 1965).
Machiavelli, Niccolo, *Discourses on the First Decade of Titus Livius* (London, 1883).
———, *Discorsi di Nicolo Machiavelli Fiorentino, sopra la prima deca di Tito Livio* (Venegia, 1554).
Maizeroy, M. Joly de, *Cours de tactique théorique, pratique, et historique, qui applique les exemples aux préceptes, développe les maximes des plus habiles généraux, & rapporte les faits les plus intéressans & les plus utiles, avec les descriptions de plusieurs batailles anciennes* (vol. 1, Paris, 1766).

Mann, J.C., 'The Frontiers of the Principate', in H. Temporini, ed., *Aufstieg und Niedergang der romischen Welt* 2, 1 (1974), pp. 508–33.

Marquardt, Joachim, and Theodor Mommsen, *Handbuch der römischen Alterthümer, römische Staatsverwaltung von Joachim Marquardt, zweiter Band* (vol. 5, Leipzig, 1884).

Marquardt, Karl Joachim, and Theodor Mommsen, *Handbuch der römischen Alterthümer, Staatsverwaltung von J. Marquardt II* (vol. 5, Leipzig, 1876).

Marx, Anthony W., *Faith in Nation: Exclusionary Origins of Nationalism* (Oxford, 2003).

Massa-Pairault, F.H., 'Notes sur le problème du citoyen en armes : cité romaine et cité étrusque', in A.-M. Adam, and A. Rouveret, ed., *Guerre et sociétés en Italie aux Ve et IVe siècles avant J-C* (Paris, 1986), pp. 29–50.

Mattern, Susan. P., *Rome and the Enemy: Imperial Strategy in the Principate* (Berkeley, 1999).

Matthew, Christopher, 'The Battle of Vercellae and the Alteration of the Heavy Javelin (Pilum) by Gaius Marius - 101 BC', *Antichthon*, 44 (2010), pp. 50–66.

Matthew, Christopher Anthony, *On the Wings of Eagles: The Reforms of Gaius Marius and the Creation of Rome's First Professional Soldiers* (Newcastle upon Tyne, 2010).

Mauricius, Imperator, edited by Ernst Gamillscheg, and George T. Dennis, eds., *Das Strategikon des Maurikios* (Wien, 1981).

Mauricius, Imperator, trans. G.T. Dennis, *Maurice's Strategikon: Handbook of Byzantine Military Strategy* (Philadelphia, 1984).

Mazzarino, S., *The End of the Ancient World* (London, 1966).

Mazzocco, Angelo, and Marc Laureys, *A New Sense of the Past: The Scholarship of Biondo Flavio (1392–1463)* (Leuven, 2016).

McArthur, A.A., 'Should Roman Soldiers be called "Professional" prior to Augustus?', *Journal of Military History*, 85:1 (2021), pp. 9–26.

McCall, Jeremiah, 'The Manipular Army System and Command Decisions in the Second Century', in Jeremy Armstrong, and Michael P. Fronda, eds., *Romans at War: Soldiers, Citizens, and Society in the Roman Republic* (London, 2020), pp. 201–31.

McCall, Jeremiah B., *The Cavalry of the Roman Republic: Cavalry Combat and Elite Reputations in the Middle and Late Republic* (London, 2002).

McNeill, W.H., *The Pursuit of Power* (Oxford, 1983).

Merivale, C., *History of the Romans Under the Empire* (vol. I, New York, 1863).

Michelet, Jules, *Histoire romaine, première partie : république* (vol. 1, Paris, 1834).

———, *Histoire romaine. 1ère partie : république* (vol. 2, Paris, 1843).

Millar, F., *A Study of Cassius Dio* (Oxford, 1964).

Miller, M.C.J., 'The Professionalization of the Roman Army in the Second Century B.C.', (Ph.D. thesis, Chicago, 1984).

———, 'The Principes and the so-called Camillan Reforms', *The Ancient World*, 23 (1992), pp. 59–70.

Milne, Kathryn H., 'The Republican Soldier: Historiographical Representations and Human Realities', (Ph.D. thesis, Philadelphia, 2009).

Mjøset, Lars, and Stephen Van Holde, 'Killing for the State, Dying for the Nations: An Introductory Essay on the Life Cycle of Conscription onto Europe's Armed Forces', in Lars. Mjøset, and Stephen. Van Holde, eds., *The Comparative Study of Conscription in the Armed Forces* (Oxford, 2002), pp. 3–94.

Momigliano, A., 'Ancient History and the Antiquarian', *Journal of the Warburg and Courtauld Institutes*, 13 (1950), pp. 285–315.

———, *Essays in Ancient and Modern Historiography* (Oxford, 1977).

Mommsen, T., *The History of Rome, Volume III* (vol. III, London, 1863).

———, *A History of Rome Under the Emperors, Based on the Lecture Notes of Sebastian and Paul Hensel, 1882–6* (London, 1996).

Mommsen, Theodor, and Joachim Marquardt, *Römisches Staatsrecht* (vol. 1, Graz, 1952).

Mommsen, Theodor, *Römische Geschichte, erster Band bis zur Schlacht von Pydna* (vol. I, Berlin, 1856).

———, *Römische Geschichte, zweiter Band von der Schlacht bei Pydna bis auf Sullas Tod* (vol. II, Berlin, 1857).

———, *The History of Rome, Volume IV* (vol. IV, New York, 1871).

Mommsen, Theodor, *Römische Geschichte: Die Provinzen von Caesar bis Diocletian* (vol. V, Berlin, 1885).

Montecuccoli, Raimondo, *Mémoires de Montecucculi généralissime des troupes de l'emperor, divisez en trois livres. 1. de l'art militaire en général. 2. de la guerre contre le turc. 3. relation de la campagne de 1664.* (Strasbourg, 1735).

Myers, Philip Van Ness, *A History of Rome* (London, 1904).

Nafziger, George F., *Imperial Bayonets: Tactics of the Napoleonic Battery, Battalion, and Brigade as Found in Contemporary Regulations* (London, 1996).

Napoli, Joëlle, 'Rome et le recrutement de mercenaires', *Revue historique des armées*, (2010), pp. 68–77.

Newman, F.W., *Regal Rome: An Introduction to Roman History* (London, 1852).

Nicolet, C., *The World of the Citizen in Republican Rome* (London, 1980).

Niebuhr, B.G., *Römische Geschichte, erster Theil* (Berlin, 1828).

———, *Römische Geschichte, zweiter Theil* (Berlin, 1812).

———, *Römische Geschichte, dritter Theil* (Berlin, 1832).

———, *Römische Geschichte, vierter Theil* (Jena, 1844).

Nisbet, R.A., *History of the Idea of Progress* (1980).

Nisbet, R.G.M., 'Aeneas Imperator: Roman Generalship in an Epic Context', *Oxford Readings in Virgil's Aeneid*, 18 (1990), pp. 378–89.

Oakley, S.P., 'Single Combat in the Roman Republic', *The Classical Quarterly*, New Series 35 (1985), pp. 392–410.

Paris, F.A., *Traite de tactique appliquée : élaboré d'après le programme prescrit pour les écoles royales de guerre allemandes* (Paris, 1873).

Parker, H.M.D., *The Roman Legions* (London, 1928).

———, *The Roman Legions* (London, 1958).

Patterson, John, 'Military Organisation and Social Change in the Later Roman Republic', in John Rich, and Graham Shipley, eds., *War and Society in the Roman World* (London, 1993), pp. 92-112.

Pearson, Elizabeth H., *Exploring the Mid-Republican Origins of Roman Military Administration: With Stylus and Spear* (London, 2021).

Pelham, H.F., *Outlines of Roman History* (London, 1893).

Phang, Sara Elise, Military Service. *Ideologies of Discipline in the Late Republic and Early Principate* (Cambridge, 2008).

———, *The Marriage of Roman Soldiers (13 B.C.- A.D. 235): Law and Family in the Imperial Army* (Boston (Mass.), 2001).

Pitassi, Michael., *The Roman Navy: Ships, Men & Warfare 350 BC–AD 475* (Barnsley, 2012).

Poe, Bryce, II, 'British Army Reforms 1902–1914', *Military Affairs*, 31 (1967), pp. 131–38.

Potter, D.S., and C. Damon, 'The Senatus Consultum de Cn Pisone Patre', *American Journal of Philology*, 120 (1999), pp. 13–41.
Raaflaub, K.A., 'Between Myth and History: Rome's Rise from Village to Empire (the Eight Century to 264)', in Nathan Rosenstein, and Robert Morstein-Marx, eds., *A Companion to the Roman Republic* (Malden, 2006)), pp. 125–46.
Raaflaub, K.A., L.J.F. Keppie, and W.S. Hanson, 'The Political Significance of Augustus' Military Reforms', *Roman Frontier Studies* 1979 XII (1980), pp. 1005–25.
Rambaud, R.M., *L'art de la déformation historique dans les commentaires de César* (Paris, 1966).
Rathmann, Michael, 'The *Tabula Peutingeriana* and Antique Cartography', in Serena Bianchetti, Michele R. Cataudella, and Hans-Joachim Gehrke, eds., *Brill's Companion to Ancient Geography: The Inhabited World in Greek and Roman Tradition* (Leiden, 2016), pp. 337–62.
Raudzens, George, 'War-Winning Weapons: The Measurement of Technological Determinism in Military History', *Journal of Military History*, 54 (1990), pp. 403–34.
Ravenel, J., *Oeuvres complètes de Montesquieu* (Paris, 1834).
Rawlings, L., 'Army and Battle During the Conquest of Italy (350–264 BC)', in P. Erdkamp, ed., A *Companion to the Roman Army* (Oxford, 2007), pp. 45–62.
Rawson, E., 'The Literary Sources for the Pre-Marian Army', *Papers of the British School at Rome*, 39 (1971), pp. 13–31.
Reeve, M. D., ed., *Vegetius Epitoma Rei Militaris* (Oxford, 2004).
Reinhold. M., 'In praise of Cassius Dio', *L'Antiquité Classique*, 55 (1986), pp. 213–22.
Riccobono, S., G. Baviera, C. Ferrini, G. Furlani, and V. Arangio-Ruiz, *Fontes Iuris Romani Antejustiniani* (3 vols, Florence, 1940).
Rich, John, 'The supposed Roman manpower shortage of the later second century BC', *Historia*, 32 (1983), pp. 287–331.
———, 'Warfare and the Army in Early Rome', in Paul Erdkamp, Anthony Birley, William Broadhead, Lukas De Blois, Pierre Cagniart, Hugh Elton, Gary Forsythe, Luuk De Ligt, and Clifford Ando, eds., *A Companion to the Roman Army* (Oxford, 2007).
———, 'Augustus and the spolia opima', *Chiron*, 26 (1996), pp. 85–127.
———, and Graham Shipley, eds., *War and Society in the Roman World* (2002).
Richardson, J.S., *Hispaniae: Spain and the Development of Roman Imperialism, 218–82 B.C.* (Cambridge, 1986).
Riedel, Meredith L.D., "God Has Sent Thunder'. Ideological Distinctives of Middle Byzantine Military Manuals', in James T. Chlup, and Conor Whately, eds., *Greek and Roman Military Manuals: Genre and History* (London, 2021), pp. 245–264.
Rocquancourt, Jean Thomas, *Cours complet d'art et d'histoire militaires: ouvrage dogmatique, littéraire et philosophique à l'usage des élèves de l'école royale spéciale militaire* (vol. 1, Paris, 1840).
Rodgers, Nigel, *The Roman Army: Legions, Wars and Campaigns: A Military History of the World's First Superpower from the Rise of the Republic and the Might of the Empire to the Fall of the West* (London, 2005).
———, *Die römische Armee: die Legionen der antiken Weltmacht und ihre Feldzüge* (Tosa, 2011).
Rosenstein, N., 'Military Command, Political Power, and the Republican Elite', in P. Erdkamp, ed., *A Companion to the Roman Army* (Oxford, 2007), pp. 132–48.
Rosenstein, Nathan S., 'Integration and Armies in the Middle Republic', in Saskia T. Roselaar, ed., *Processes of Integration and Identity Formation in the Roman Republic* (Boston, Mass., 2012), pp. 71–83.

Rosenstein, Nathan Stewart, *Rome at War: Farms, Families, and Death in the Middle Republic* (Chapel Hill, N.C., 2004).
Rostovtzeff, M.I., *A History of the Ancient World* (vol. II, London, 1945).
Roth, Jonathan P., *The Logistics of the Roman Army at War: (264 B.C.-A.D. 235)* (Boston, Mass., 1999).
——, *Roman Warfare* (Cambridge, 2009).
Rudow, Alexander, *Die römische Armee: Organisation, Ausrüstung, Eroberungen* (Rheinbach, 2015).
Rüstow, Wilhelm, *Der Krieg und seine Mittel; eine allgemein fassliche Darstellung der ganzen Kriegskunst* (Leipzig, 1856).
Rüstow, Wilhelm, *Heerwesen und Kriegführung C. Julius Cäsars* (Nordhausen, 1862).
Sabin, Philip, 'The Face of Roman Battle', *The Journal of Roman Studies*, 90 (2000), pp. 1–17.
Saddington, D.B., *The Development of the Auxiliary Forces from Caesar to Vespasian* (Harare, 1982).
Sage, Michael M., *The Army of the Roman Republic: From the Regal Period to the Army of Julius Caesar* (Barnsley, 2018).
Salvadori, M., *The Liberal Heresy* (London, 1977).
Santosuosso, A., *Storming the Heavens* (London, 2004).
Sanz, Fernando Quesada, 'Not So Different: Individual Fighting Techniques and Battle Tactics of Roman and Iberian Armies Within the Framework of Warfare in the Hellenistic Age', in P. François, P. Moret, and S. Péré-Noguès, eds., *L'Hellénisation en méditerranée occidentale au temps des guerres puniques*. Actes du Colloque International de Toulouse, 31 mars-2 avril 2005 *Pallas* 70 (2006), pp. 245–63.
Sarkesian, Sam C., and Connor, Robert E. Jr, *The US Military Profession into the Twenty-First Century: War, Peace and Politics* (Oxford, 2006).
Sarkissian, J., 'The Idea of Imperium in Aeneid 1, 50–296', *The Augustan Age*, 4 (1985), pp. 51–6.
Scapini, Marianna, "Literary Archetypes for the Regal Period', in Bernard Mineo, ed., *A Companion to Livy* (2015), pp. 274–85.
Schlieffen, Alfred von, and Hugo Friedrich von Freytag-Loringhoven, *Cannae mit einer Auswahl von Aufsätzen und Reden des Feldmarschalls sowie einer Einführung und Lebensbeschreibung von General der Infanterie Freiherr von Freytag-Loringhoven* (Berlin, 1925).
Schmitthenner, W., 'The Armies of the Triumval Period: A Study of the Origins of the Roman Imperial Legions', (D.Phil. thesis, Oxford, 1958).
Schmitz, Leonard, *Lectures on the History of Rome: From the Earliest Times to the Fall of the Western Empire by R.G. Niebuhr* (vol. I, London, 1850).
Schneider, Hans-Christian., *Das Problem der Veteranenversorgung in der späteren römischen Republik* (Bonn, 1977).
Schwegler, Albert, Römische Geschichte im Zeitalter der Könige (Tübingen, 1869).
——, *Römische Geschichte im Zeitalter des Kampfs der Stände* (vol. 2, Tübingen, 1872).
——, *Römische Geschichte von Gallischen Brande Roms bis zum ersten samniter* Kreige (vol. 4, Tübingen, 1873).
Scullard, H. H., *A History of the Roman World From 753 to 146 B.C.* (London, 1951).
——, *From the Gracchi to Nero* (London, 1959).
——, *Roman Politics 220–150 B.C.* (Oxford, 1973).
——, *From the Gracchi to Nero* (London, 2010).

Sekunda, N., S. Northwood, and R. Hook, *Early Roman Armies* (London, 1995).
Sekunda, Nicholas, 'Military Forces. A. Land Forces B. Naval Forces', in Philip Sabin, Hans van Wees, and Michael J. Whitby, eds., *The Cambridge History of Greek and Roman Warfare: Volume I Greece, the Hellenistic World and the Rise of Rome* (Cambridge, 2007), pp. 325–67.
Sherk, Robert K., and Hildegard Temporini, 'Roman geographical exploration and military maps', *Politische Geschichte (Allgemeines)*, (1974), pp. 534–62.
Shotter, David, *Augustus Caesar* (London, 1991).
Sim, David, 'The manufacture of disposable weapons for the Roman army', *Journal of Roman Military Equipment Studies*, 3 (1992), pp. 105–19.
Sim, David, and Jaime Kaminski, *Roman Imperial Armour: The Production of Early Imperial Military Armour* (Oxford, 2012).
Simkins, Michael, and Ronald Embleton, *Die römische Armee: Von Caesar bis Constantin (44 V. Chr.-333 N. Chr.)* (Sankt Augustin, 2005).
Simkins, Peter, *Kitchener's Army: The Raising of the New Armies, 1914–16* (Manchester, 1988).
Smith, R. E., *Service in the Post-Marian Roman Army* (Manchester, 1958).
Smith, W. G., *Dictionary of Greek and Roman Biography and Mythology* (vol. III, London, 1849).
Southern, Pat, *The Roman Army: A Social and Institutional History* (Oxford; New York, 2007).
Southern, Pat, and Karen R. Dixon, *The Late Roman Army* (London, 1996).
Starr, Chester G., *The Roman Imperial Navy, 31 B.C.-A.D. 324* (Ithaca, 1941).
Steinwender, Theodor, *Die Marschordnung des römischen Heeres zur Zeit der Manipularstellung* (Danzig, 1907).
Storch, R. H., 'The Archaic Greek Phalanx 750–650 B.C.', *The Ancient History Bulletin*, 12 (1998), pp. 1–7.
Streit, Pierre, *L'armée romaine* (Gollion, 2012).
Strobel, K., 'Strategy and Army Structure Between Septimius Severus and Constantine the Great', in P. Erdkamp, ed., *A Companion to the Roman Army* (Oxford, 2007), pp. 267–86.
Sumner, Graham, *Die römische Armee: Bewaffnung und Ausrüstung* (Stuttgart, 2007).
Swain, Col USA (Ret) Richard, *The Obligations of Military Professionalism: Service Unsullied by Partisanship* (Washington, 2010).
Swan, P. M., 'How Cassius Dio composed his Augustan books: four studies.', *Aufstieg und Niedergang der romischen Welt*, Principät II (1997), pp. 2524–2558.
Syme, R., 'Some Notes in the Legions under Augustus, *The Journal of Roman Studies*, 23 (1933), pp. 14–33.
———, 'The Spanish War of Augustus (26–25 BC)', *The American Journal of Philology*, 55 (1934), pp. 293–317.
———, *The Roman Revolution* (London, 1939).
———, 'The Origin of the Veranii', *The Classical Quarterly, New Series*, 7 (1957), pp. 123–5.
Tan, James, 'The *Dilectus-Tributum* System and the Settlement of Fourth Century Italy.', in Jeremy Armstrong, and Michael P. Fronda, eds., *Romans at War: Soldiers, Citizens, and Society in the Roman Republic* (London, 2020), pp. 52–75.
Taylor, Michael J., 'The battle scene on Aemilius Paullus's Pydna monument: a reevaluation', *Hesperia: The Journal of the American School of Classical Studies at Athens*, 85 (2016), pp. 559–76.
———, 'Visual evidence for Roman infantry tactics', *Memoirs of the American Academy in Rome*, 59 (2014), pp. 103–20.

———, 'Etruscan identity and service in the Roman army: 300–100 B.C.E.', *American Journal of Archaeology: The Journal of the Archaeological Institute of America*, 121 (2017), pp. 219–36.

Todière, M., *Sommaire d'un cours complet d'histoire romaine* (Tours, 1846).

Juliusz Tomczak, 'Roman Military Equipment in the 4th Century BC: Pilum, Scutum and the Introduction of Manipular Tactics', *Acta Universitatis Lodziensis Folia Archaeologica*, 29 (2012), pp. 38–65.

D.J.B. Trim, ed., *The Chivalric Ethos and the Development of Military Professionalism* (Leiden, 2003).

A.V. Tucker, 'Army and Society in England 1870–1900: a Reassessment of the Cardwell Reforms, *The Journal of British Studies*, 2 (1962), pp. 110–41.

van Crefeld, M., *The Art of War: War and Military Thought* (London, 2000).

Veith, Georg, *Geschichte der Feldzüge C. Julius Caesars* (Wien, 1906).

von Arnim, H., 'Ineditum Vaticanum', *Hermes*, 27 (1892), pp. 118–30.

von Decker, C., *De la tactique des trois armes, infanterie, cavalerie, artillerie, isolées et réunies dans l'esprit de la nouvelle guerre cours fait à l'école militaire de berlin par c. de Decker: contenant la tactique de chaque arme isolée* (vol. 1, Brussels, 1836).

von Domaszewski, A., *Die Rangordnung des Römischen Heeres* (Bonn, 1908).

von Hardegg, J., *Anleitung zum Studium der Kriegsgeschichte* (vol. 1, Darmstadt, 1868).

von Schlieffen, trans. Anon., General Field Marshall Alfred, *Cannae* (Fort Leavenworth, 1931).

Walbank, F. W., *A Historical Commentary on Polybius, Volume 1* (vol. I, Oxford, 1957).

———, *Polybius, Rome and the Hellenistic World: Essays and Reflections* (Cambridge, 2002).

F. W. Walbank, A. E. Astin, M. W. Frederiksen, and R. M. Ogilvie, eds., *The Cambridge Ancient History Volume VII Part 2: The Rise of Rome to 220B.C.* (Cambridge, 1989).

Walbank, F.W., *Polybius* (Berkeley, 1972).

Wallace-Hadrill, Andrew, 'Mutatas Formas: The Augustan Transformation of Roman Knowledge', in Karl Galinsky, ed., *The Cambridge Companion to the Age of Augustus* (Cambridge, 2005), pp. 55–84.

Walsh, P.G., *Livy* (Cambridge, 1961).

Watson, G. R., *The Roman Soldier* (London, 1969).

Webster, G., *The Roman Imperial Army of the First and Second Centuries A.D.* (Norman, 1998).

Webster, H., *Ancient History* (Boston, 1913).

Wells, C. M., *The Roman Empire* (London, 1984).

Wells, C.M., *The German Policy of Augustus: An Examination of the Archaeological Evidence* (Oxford, 1972).

Wells, P.S., *The Barbarians Speak: How the Conquered Peoples of Europe Shaped the Roman Empire* (Princeton, 1999).

Wells, Peter S., *The Battle That Stopped Rome: Emperor Augustus, Arminius, and the Slaughter of the Legions in the Teutoburg Forest* (New York, 2003).

Whately, Conor, *An Introduction to the Roman Military: From Marius (100 BCE) to Theodosius II (450 CE)* (Hoboken, 2021).

———, 'Military Manuals from Aeneas Tacitus to Maurice. Origins, Scholarship, Genre, Audience and History', in James T. Chlup, and Conor Whately, eds., *Greek and Roman Military Manuals: Genre and History* (London, 2021).

Wheeler, E. L., 'Methodological Limits and the Mirage of Roman Strategy: Part 1', *The Journal of Military History*, 57 (1993), pp. 7–41.

———, 'Methodological Limits and the Mirage of Roman Strategy: Part 2', *The Journal of Military History*, 57 (1993), pp. 215–40.
Wheeler, Everett L, 'The Army and the Limes in the East', in P. Erdkamp, ed., *A Companion to the Roman Army* (Oxford, 2007), pp. 235–66.
Wheeler, Everett L., *Stratagem and the Vocabulary of Military Trickery* (New York, 1988).
Whittaker, C.R., *Rome and Its Frontiers: The Dynamics of Empire* (London, 2004).
———, *The Frontiers of the Roman Empire: A Social and Economic Study* (Baltimore, 1994).
Williams, Sarah, and Alan Apperley, 'Public Relations and Discourses of Professionalism', in A. Rogojinaru, and S. Wolstenholme, eds., *Current Trends in International Public Relations* (Bucharest, 2009), pp. 1–20.
Wilms, Albert, *Die Schlacht bei Cannae* (Hamburg, 1895).
Wolff, Catherine, *L'armée romaine : une armée modèle* (Paris, 2012).
Woodward, D., *The Armies of the World 1854–1914* (London, 1978).
Young, Thomas-Durell, 'Military Professionalism in a Democracy', in Thomas C. Bruneau, and D. Tollefson Scott, eds., *Who Guards the Guardians and How* (Austin, 2006), pp. 17–34.
Zhmodikov, Alexander, 'Roman Republican Heavy Infantrymen in Battle (IV-II Centuries B.C.)', *Historia: Zeitschrift für Alte Geschichte*, 49 (2000), pp. 67–78.
Zipfel, Lasse, Augustus Heeresreform. *Das Heer als Grundlage des römischen Imperiums* (München, 2015).

Index

πολιτεία 86
πολιτέια 4
στρατιῶται ἀθάνατοι, 30

Actium 121, 133, 135, 138, 141, 152, 156
  Battle of 56, 66
Aelian 6, 48, 89, 92
Aemilius Paullus 47, 96, 110, 126
Aeneas Tacticus 48
Aerarium Militare 134, 140, 149, 151, 153, 154
Africa 19, 54, 103, 105, 106, 107, 118, 128, 129, 134, 142
Agmen 28
Altar of Domitius Ahenobarbus 110, 125
Ammianus xii, xiv, 29, 37, 49
Antesignarii 122
Antonius, M. 119, 120, 141
Appian 108, 114, 118, 119, 121
Armenia 135, 144
Arrian xii, 3, 48, 49, 52, 89, 92, 101
Asclepiodotus 37, 48, 89, 91, 92
Assidui 77
Athesis 116
Augusta Emerita 144, 145
Augusta Praetoria 144
Augustan 31, 35
Augustine 37, 39
Augustus xi, xii, xiii, 4, 6, 8, 18, 30, 32, 33, 35, 36, 41, 48, 49, 54, 56, 59, 61, 64, 66, 68, 73, 74, 76, 82, 84, 86, 89, 110, 111, 113, 132, 133, 134
Aulus Hirtius 119, 122
Auspicium 84
Auxilia sociorum 124, 136, 141

Baecula 105
Becker, A. 16
Bell, M.J.V. 112, 115
Berenhorst, Georg Heinrich von 11
Bibracte 125
Biondo Flavio 5

Booty 85, 95, 102, 103, 105, 107, 141, 146, 147, 149, 153, 154
Byzantine 5, 32

Cadenced marching 99
Caecilius Metellus Numidicus, Q. 121
Caetrati and *scutatae* 129
Cagniart 64
Camillan 35, 80
Cannae 20, 22, 94, 98, 100, 105, 106, 128
Cantabri and Astures 144
Capite censi 112, 113, 114, 117, 118, 153
Carfulenus, Decimus 119
Carniart 35
Carrhae 143
Carthage 40, 84, 98, 107
Cato 40, 48, 76, 107
Cavalry 8, 13, 35, 48, 49, 50, 54, 58, 75, 87, 90, 91, 92, 93, 94, 97, 107, 108, 110, 115, 116, 123, 124, 127, 128
Censors 90, 124
Centurions 21, 62, 66, 84, 90, 92, 97, 119, 122, 123, 127, 130
Certosa Situla x, 77
Chain mail, more correctly ring mail 126
Christian thought 38
Cilicia 114
Cimbri and Teutones 121
Cimbric War 115
Cincius 48
Civil war 119
Clades variana 151, 154
Clan warfare 79
Clarke, Napoleon's Minister of War 27
Claudius Nero 100, 108
Claudius Quadrigarius 74
Clientela 106
cohort/s 17, 21, 61, 64, 98, 99, 110, 111, 112, 114, 115, 120, 123, 127, 128, 135, 136, 155
Comitia Centuriata 34, 75
Conditions of service 52, 137, 138
Constitutiones 48, 137

constitutiones 50, 52, 137, 139
contingent adaptation xiii, 80, 83, 93, 95, 102, 104, 117, 118, 121, 129, 137, 153, 155
contingent decision 43, 45, 52, 53, 54, 55, 100, 101, 102, 103, 104, 106, 107, 109, 124, 150

Dacians 140, 148, 151
Dalmatia/ns 140, 145, 147, 148
Danube 135, 140, 145, 147, 148, 150
Deiotarus, king of Galatia 120
De la Barre Duparcq 15
De la Chauvelays 18
Dilectus 100, 119, 124
Dionysius of Halicarnassus 4, 73, 75, 80, 82, 86, 128
Disciplina 68, 69, 138, 140
Drusus, Julius Caesar 151
Drusus, Nero Claudius 145, 147, 148, 149

Egypt 87, 98, 120, 125, 140, 141, 142, 144, 152
Eighteenth-Century Enlightenment 10
Elephants 95, 97, 108
Encampment practices 48, 49, 93, 94
Equus publicus 116
Evocati 53, 122
exercitus 28, 46, 76
Expediti 53, 54, 122
Extraordinarii 93

Fabius Maximus 7, 106
Fabius Pictor 74, 76
Flamininus 93, 95, 109
Flaminius 81, 100
French Revolution 12, 26
Frontinus 6, 40, 48, 51, 52
Fustel de Coulanges 16, 111

Gauls 81, 82, 151
Germanicus, Julius Caesar 151, 154
Germany 11, 17, 19, 21, 23, 27, 139, 140, 143, 145, 147, 148, 149, 150, 152, 154
Gladius hispaniensis 125
Greaves 82, 126

Hadrian, Imperator P. Aelius Hadrianus 19, 48, 50, 51, 138
Hanson, V.D. 22, 101
Hasdrubal 105, 107
Hastae 100, 123
Hastati 82, 90, 91, 95, 96, 99, 108, 123, 124
Hellenistic warfare 97

Helvetii 125
Herodian 30, 69
Hispania Tarraconensis 143
Hoplites 75, 77, 78, 83, 101

Ilipa 105, 114
Illyricum 134, 140, 142, 143, 150
Imperium 4, 45, 46, 55, 68, 84, 105, 137, 140, 143
Ineditum Vaticanum 81
Italica 106

Joly de Maïzeroy 9
Jugurtha 99, 115, 121, 130
Julius Africanus 48, 53
Junkekmann, M. 133

Kalkriese 151
Kant 11
Kertész, I. 112
Kriegswesen 19, 23
Kubitschek, W. 19, 61
Kynoskephalai 109

Ladner, G.B. 38
La Turbie monument 146
Lawgiver 76
Le Beau 17
Le Bohec, Y. 133
Legati 107, 137, 143, 145, 151
Legio 76, 120, 143
Legio I Augusta 143
Legio II Augusta 143
Legio IV Macedonica 143
Legio V Alaudae 120, 143
Legio VI Victrix 143
Legio IX Hispana 143
Legio X Gemina 143
Legio XXII Deiotariana 120
Leighton, R.F. 61
Lendon, J.E. 68, 69
Levis armaturae 122
Ligustinus 63, 65, 66, 67, 90
Logistics 43, 44, 57, 103, 104, 156
Lucullus, L. Licinius 112, 115, 131

Macedonian phalanx 4, 96, 129
Macedonians 10, 37, 88, 89, 94, 96, 99, 109, 112, 120, 129
Maecenas 30, 69
Maniple 37, 82, 90, 97, 99, 110, 112, 114, 123
Manipularis 123
Marcellus 100

Marcomanni 150
Marian 35, 61, 65, 68, 110, 111, 115
Marian reforms 65, 68
Marius, C. 117
Mattern, S.P. 133
Maurice de Saxe 8
Mauricius, emperor 5, 48
Mercenaries 18, 57, 94, 102, 107, 108
Metaurus 100, 108, 114
Militärwesen 17
milites 28, 66
militia 28, 58, 62, 64, 113
Minucius Lorarius of Legio Tertia 127
Moesia 151
Montesquieu 59
Mutiny 53, 105, 106

Napoleon 12, 14, 27, 135
Napoleonic Wars 17, 23, 57, 67
National armies xi, 25, 27, 28, 31
Niebuhr, B. 16, 34, 60
Nova Carthago 102, 104
Numantia 126
Numa Pompilius 75
Numidia 115

Octavian 65, 135, 138, 141, 152, 156
Onasander 48, 50, 52
Orchomenos 115

Pannonia 143, 145, 147, 148, 149, 150, 152, 153
Pannonian revolt 150
Pannonians 140, 150
Panoply 83, 91, 122
Pansa Caetronianus, C. Vibius 119
Paris, Major General Friedrich August 13, 20
Parma 91
Parthians 140, 143, 144
Peace of Westphalia 7, 26
Peysegur 9
Phang, S.E. 137, 138
Philip V 98, 109
Phraates, king of Parthia 144
Pilum 75, 82, 83, 117, 125, 131
Pleminius 107
Plutarch 81, 101, 107, 116, 118
Polyaenus 48, 52, 53
Polybius xi, xii, xiii, xiv, 4, 6, 7, 15, 21, 35, 37, 40, 67, 71, 81, 82, 84, 86, 87, 88, 89, 90, 91, 92, 93, 94, 95, 96, 97, 98, 100, 101, 104, 108, 110, 111, 114, 115, 117, 118, 120, 121, 123, 124, 125, 126, 127, 128, 129, 130, 138, 146, 150, 155, 156
Pomerium 84, 142
Pompeius, Gn. Magnus 46, 54, 63, 116, 120, 123, 126, 128, 129, 135, 139, 143, 146
Potestas 84
Praefecti 93
Praetorian Guard 135
Primus pilus 127
Principes 12, 82, 90, 91, 95, 99, 108, 123, 124
professional 12, 15, 17, 19, 21, 23, 28, 47, 56, 58, 59, 60, 61, 64, 67, 69, 111, 113, 117, 132, 136, 138, 156
professionalism 56, 57, 58, 59, 60, 61, 63, 64, 65, 66, 67, 68, 69, 73, 110, 111, 113, 139, 155
Proletarianisation 63
Provincia 46, 84, 142, 148
Prussians 27
Publius Valerius Publicola 78
Pydna 66, 91, 96, 98
Pyrrhus 14, 80, 96

Recruitment 5, 100, 101, 112, 117
reform xii, 32, 33, 34, 35, 36, 37, 38, 39, 41, 42, 61, 63, 73, 74, 80, 83, 110, 111, 132, 133, 139, 140, 155
Reform Acts 39
Renaissance 5, 16, 20, 38
Renieblas and Numantia 94
Res Gestae Divi Augusti 140
RGDA Divi Augusti, See *Res Gestae Divi Augusti*
Rhaetia 147
Rhine 147, 148, 151, 153
Ritterling 19
Roman army xi, xiii, 3, 6, 16, 17, 18, 19, 21, 23, 24, 25, 26, 27, 28, 31, 32, 33, 34, 35, 41, 54, 55, 56, 61, 64, 68, 69, 77, 80, 98, 105, 116, 132, 133, 137, 155
Roman deployment 95, 108

Sacramentum 128, 140
Saddington, D. 136
Salassi 143, 144
Samnites 82, 94
Satricum 78
Scipio Aemilianus 34, 87
Scipio Africanus 53, 61, 95, 101, 102, 103, 108, 113, 135
Scutum 75, 82, 83
Second Sophistic 89
Senate 43, 45, 100, 102, 103, 106, 117, 122, 146, 149

Septimus, L. Marcius 102
Servian 35, 74, 78, 79
Servius Tullius 32, 35, 74, 77, 79
Sicily 102, 106, 107
Social War 111, 117, 120, 121, 124, 131, 136
Socii, i.e. allies 85, 112, 141
Sociorum auxilia 122
Spain 6, 54, 55, 61, 65, 90, 91, 94, 98, 102, 103, 104, 105, 107, 114, 115, 118, 120, 123, 126, 127, 129, 140, 141, 143, 149, 152
Spanish swords 104
Spolia opima 142
Standing armies xii, 30, 61, 64–6, 97, 99, 113–14, 134–6
Stipendia 61, 68, 80, 85, 116, 150, 154
Stipendium 80
stratagems 48, 52, 53, 54, 55, 101, 108
Suetonius, C. Tranquillus 139, 140, 145, 149, 150, 153
Supplementum 114, 117
Syphax 105, 107
Syracuse 106
Syria 88, 98, 114, 135

Tabula Heracleensis 116
Taylor, M.J. 112
Temple of Janus 144, 148
Tenney Frank 61
Terentius Varro, A. 143
Thucydides xii, xiv, 40, 43, 88, 92, 101, 156
Training 5, 10, 43, 44, 48, 50, 51, 52, 57, 58, 62, 83, 84, 97, 104, 105, 107, 120, 138
Trajan, Imperator M. Ulpius Traianus xii, 17, 48, 50, 138
Trajan's column 17
Triarii 82, 90, 91, 99, 100, 108, 123
Tribuni militum 67, 84
Tributum 80, 85

Tullian 35, 78

Urso (mod. Orsuna) 126

Valerius Maximus 36, 84, 120
Varus, P. Quinctilius x, 143, 151
Veii 62, 79, 80
Velites 90, 91, 99, 101, 108, 110, 112, 115, 121, 122, 129
Velleius Paterculus, C. 28, 49, 143, 153, 154
Vennonius 76
Veterans 25, 108, 109, 120, 122, 144, 151, 152, 154
Virtus 47, 69, 114, 130
Volunteers 66, 100, 106, 107, 111, 118
Von Berenhorst 11, 16
Von Berneck 20
Von Clausewitz 10, 13
Von François ix, 22
Von Hardegg 20
Von Moltke 20
Von Schlieffen 20, 22

Whately, Connor 133

Zama 14, 95, 108

Dear Reader,

We hope you have enjoyed this book, but why not share your views on social media? You can also follow our pages to see more about our other products: facebook.com/penandswordbooks or follow us on Twitter @penswordbooks

You can also view our products at www.pen-and-sword.co.uk (UK and ROW) or www.penandswordbooks.com (North America).

To keep up to date with our latest releases and online catalogues, please sign up to our newsletter at: www.pen-and-sword.co.uk/newsletter

If you would like a printed catalogue with our latest books, then please email: enquiries@pen-and-sword.co.uk or telephone: 01226 734555 (UK and ROW) or email: uspen-and-sword@casematepublishers.com or telephone: (610) 853-9131 (North America).

We respect your privacy and we will only use personal information to send you information about our products.

Thank you!